Table of Contents

Guided Explorations in General Chemistry

SECOND EDITION

David M. Hanson
SUNY Stony Brook

BROOKS/COLE
CENGAGE Learning

Australia • Brazil • Japan • Korea • Mexico • Singapore • Spain • United Kingdom • United States

For product information and technology assistance, contact us at **Cengage Learning Customer & Sales Support, 1-800-354-9706**

For permission to use material from this text or product, submit all requests online at **www.cengage.com/permissions** Further permissions questions can be emailed to **permissionrequest@cengage.com**

ISBN-13: 978-1-4390-4965-5
ISBN-10: 1-4390-4965-3

Brooks/Cole
20 Davis Drive
Belmont, CA 94002-3098
USA

Cengage Learning is a leading provider of customized learning solutions with office locations around the globe, including Singapore, the United Kingdom, Australia, Mexico, Brazil, and Japan. Locate your local office at: **www.cengage.com/global**

Cengage Learning products are represented in Canada by Nelson Education, Ltd.

To learn more about Brooks/Cole, visit **www.cengage.com/brookscole**

Purchase any of our products at your local college store or at our preferred online store **www.cengagebrain.com**

Printed in the United States of America
1 2 3 4 5 6 7 14 13 12 11 10

To Instructors and Students

General Chemistry: Guided Explorations consists of activities that not only help students master the concepts and procedures in a General Chemistry course but also engage them in learning, grow performance skills, enhance conceptual understanding, and facilitate problem solving. It can be used in either large or small class meetings or as homework with a traditional textbook as a reference resource.

The design of the activities is based in research on how people learn.[1] This research also forms the foundation of our successful POGIL model for college instruction[2] and the popular 5E/7E model at the secondary level.[3] Briefly, people learn by connecting new knowledge to what they already know; following a sequence of exploration, concept formation, and application; and then reflecting on what they have learned, and how they can improve.

Each activity is built around one to three related concepts and procedures. Generally there are two activities for each chapter in a traditional General Chemistry textbook. The intent is to provide instructors with a resource that they can use to supplement and not necessarily replace their usual pedagogy and curriculum.

Activities are numbered and each begins with a title that includes a focus question. The numbering groups activities by topic; the title identifies the principal concept in the activity; and the focus question helps students connect the new ideas to what they already know. To further promote such connections, many activities include a question asking students for an opinion or to make a prediction. An introduction also motivates and sets the stage for the activity

The heart of each activity is the *Exploration*, where students are given a model to explore. The model is simply some representation of what is to be learned. It might be a diagram, a table of information, an illustrative problem, experimental data, or even some written text. Questions help guide the exploration of the model and lead to identifying and understanding the relevant concepts. The first few questions are directed. They point to relevant information in the model. Subsequent questions require that ideas be brought together and conclusions be made. These questions help students to form concepts and develop an understanding of them.

Information sections are embedded in the activity to cement what is being learned and to clarify and generalize issues or additional points.

Students then practice applying their new knowledge in *exercises* and *problems*. The exercises are straightforward. They help build confidence. Problems are more complex. Problems help integrate new knowledge with prior knowledge and develop the ability to apply knowledge in new situations. Additional problems from a textbook should be solved as well because repetition using the same concepts and procedures in different contexts is essential to learning.

At the end of each activity, a *Got It!* section helps students reflect on, assess, and document their success. If students have trouble with this section, they should read the relevant material in their textbook, and go back and do the activity over again.

Students will learn the most and have the most fun if they work together. Discussions among members of such learning teams will produce different viewpoints regarding the concepts and their use in solving problems, will identify and correct misconceptions, and will strengthen and deepen understanding of chemistry. A textbook can be used to resolve disagreements, to find answers to questions that arise, and to provide examples of problem solutions.

Success with process-oriented guided-inquiry learning requires that the instructor serve as a coach or guide-on-the-side not as the sage-on-the stage dispensing information. References describing successful classroom structures and strategies are available for both small and large classes.[2,4] Small classes are defined as up to 50 to 100 students; large classes are more than 100 students in a lecture hall.

In a large class setting, it generally is better to present the introductory material, including prerequisites, and the material in the *Information* sections in the form of mini-lectures. Otherwise too many students are likely to be lost at the beginning, not have the confidence to proceed, or be frustrated because their learning is not being supported. Essentially in a large class situation, mini-lectures, delivered only when needed, replace the facilitation provided by an instructor interacting with a group of 3 or 4 students.

The use of a student response system (clickers) is the real key for success in large classes. Selected questions from an activity are posed to the class periodically, and responses are collected. These questions set the pace, keep students on task, promote interactions among students, and provide individual accountability for the work. They also provide instant feedback to the instructor who then can support student learning through mini-lectures before frustration sets in.

(1) Hanson, D. M. "A Cognitive Model for Learning Chemistry and Solving Problems: Implications for Curriculum Design and Classroom Instruction," in *Process-Oriented Guided-Inquiry Learning*; Moog, R. S., Spencer, J. N., Eds.; American Chemical Society: Washington, DC, 2008, p 14-25.

(2) Hanson, D. M. *Instructor's Guide to Process-Oriented Guided-Inquiry Learning*; Pacific Crest: Lisle, IL, 2006.

(3) Eisenkraft, A. "Expanding the 5E Model," *Science Teacher* **2003**, *70*, 56-59.

(4) Yezierski, E. J.; Bauer, C. F.; Hunnicutt, S. S.; Hanson, D. M.; Amaral, K. E.; Schneider, J. P. "POGIL Implementation in Large Classes: Strategies for Planning, Teaching, and Management," in *Process-Oriented Guided-Inquiry Learning*; Moog, R. S., Spencer, J. N., Eds.; American Chemical Society: Washington, DC, 2008, p 60-71.

01-1. The Nature of Matter: What is that substance?

Scientists classify matter as elements, compounds, pure substances, mixtures, and solutions. They also classify changes in matter as physical changes or chemical changes.

What do you think?

1. Apply the terms element, compound, pure substance, mixture, and solution to each of the following. More than one term may apply.
   (a) helium

   (b) table salt

   (c) blood

   (d) air

   (e) tea

2. Consistent with your classifications in item 1 above, describe what each of the following terms means to you.
   (a) element

   (b) compound

   (c) pure substance

   (d) mixture

   (e) solution

Information

The use of words in science is very precise. On the street, *oily dirt* may be a good answer to the question, "What is that substance?" In the science lab, however, *oily dirt* is not a substance, it is a mixture.

A *substance*, or more explicitly *pure substance*, is any pure matter that cannot be separated into components by physical methods like picking out the pieces, evaporation, filtration, distillation, or crystallization. The composition of a pure substance is always the same. Physical methods separate the components but do not change them. Chemical methods like combustion (burning) transform the substances into other substances.

Vodka is not a pure substance. Its composition, which is the amount of ethanol and water present, can vary, and it can be separated into ethanol and water by distillation. Ethanol, on the other hand, is a pure substance. Its composition always is the same, and it cannot be converted into any other substance by physical methods, but it can be converted into carbon dioxide and water by burning, which is a chemical process.

An *element* is a pure substance that can only be decomposed into other pure substances by nuclear reactions. An *element* cannot be decomposed into two or more other pure substances by either physical or chemical methods. All of the known elements are listed on the Periodic Table. Some examples of elements are hydrogen, helium, oxygen, nitrogen, silver, gold, and lead. When elements are combined, they can form mixtures or compounds, which are described below.

A *compound* is a pure substance formed from 2 or more elements. Glucose, a sugar, is a compound. It is formed from carbon, hydrogen, and oxygen.

A *mixture* consists of two or more pure substances. A mixture can be *homogeneous* or *heterogeneous*. A homogeneous mixture is uniform: the parts are not distinguishable, like sugar dissolved in water. Air, which is made up of oxygen, nitrogen, and small amounts of other gases, also is a homogeneous mixture. A heterogeneous mixture is not uniform: the parts are distinguishable, like salad dressing or a package of white and brown rice.

The term *mixture* generally applies to a mixture that is heterogeneous, and the term *solution* is used for a mixture that is homogeneous.

A *chemical change or process* involves the transformation of one or more pure substances into one or more different pure substances. A pure substance must be an element or a compound. Iron rusting is an example of a chemical change. Iron combines with oxygen to form rust.

A *physical change or process* involves changes in pure substances, mixtures, and solutions that do not transform the pure substances present into other pure substances. Water freezing or boiling and sugar dissolving in water are examples of physical changes.

Exploration

1. How do chemists use the term *substance* in a way that probably differs from the way *substance* is used in casual conversation?

2. Two elements can form a compound or a mixture. For example, hydrogen and oxygen can combine to form water; oxygen and nitrogen can combine to form air.
   (a) Which combination is a compound, and which is a mixture?

   (b) Is this mixture homogeneous or heterogeneous? Explain.

3. What are some similarities and differences between water vapor and air, both of which are made from two elements?

4. What is the difference between a physical process and a chemical process? Provide examples of each that are not given to you in this activity.

Application

1. Examine your previous classifications of each of the following and revise them as necessary in view of the new information that you have been given. Explain your rationale for each case.

   (a) helium

   (b) table salt

   (c) blood

   (d) air

   (e) tea

2. Identify each of the following as a physical or chemical change and explain your rationale.

   (a) the fender on a car rusts

   (b) water evaporates

   (c) sugar dissolves in water

   (d) water freezes at 0 °C

   (e) food is digested

Got It!

If you have mastered this material you should be able to

1. Explain the similarities and differences between pairs of the following terms: element, compound, substance, mixture, and solution.

2. Explain the similarities and differences between a physical change and a chemical change.

2. Provide examples for each of the following terms: element, compound, substance, mixture, solution, physical change, and chemical change.

3. Classify examples that you are given as one of the following: element, compound, substance, mixture, solution, physical change, or chemical change.

4

01-2. Scientists Love to Measure: Which athlete is heavier, taller, faster?

Scientists, as well as people in general, love to measure things. Extensive statistics are kept on the performance of athletes. Scientists use measurements to identify what is happening in experiments, and to verify or reject explanations and theories. Some of the basic quantities that you will encounter are listed in Table I.

What do you think?

1. Peyton and Eli Manning are two quarterbacks. Peyton plays for the Indianapolis Colts, and Eli plays for the New York Giants. Peyton weighs 105 kg, and Eli weighs 220 lbs. Who is heavier?

2. Candace Parker and Kia Vaughn are two basketball players. They starred in the 2007 Women's NCAA basketball tournament. Candace played for Tennessee, and Kia played for Rutgers. Candace is 76 in tall, and Kia is 1.93 m tall. Who is taller?

3. Jennifer Rodriguez and Apolo Ohno are two Olympic medalists and record holders in speed skating. Jennifer's time on the 500 m long track is 37.83 s. Apolo's time on the 500 m short track is 41,518 ms. Who is faster?

Table I. International System of Units Used in Measurements (SI Units)

Physical Quantity	Unit	Abbreviation
mass	kilogram	kg
length	meter	m
volume	liter	L
time	seconds	s
amount of substance	mole	mol
temperature	kelvin centigrade or celsius	K °C

5

Prefixes are added to units to deal with very large or very small numbers. The prefixes that you need to know are listed in Table II.

Table II. Prefixes Used with Units

Prefix	Abvtn	Meaning	Example with Scientific Notation
mega	M	10^6	1,000,000 Hz = 1.0×10^6 Hz= 1 MHz
kilo	k	10^3	1,000 g = 1.0×10^3 g= 1 kg
centi	c	10^{-2}	0.01 m = 1.0×10^{-2} m = 1 cm
milli	m	10^{-3}	0.001 L = 1.0×10^{-3} L = 1 mL
micro	µ	10^{-6}	0.000001 g = 1.0×10^{-6} g = 1µg
nano	n	10^{-9}	0.000000001 m = 1.0×10^{-9} m = 1 nm
pico	p	10^{-12}	0.000000000001 m = 1.0×10^{-12} m = 1 pm
femto	f	10^{-15}	0.000000000000001 s = 1.0×10^{15} s = 1 fs

Exploration - 1

1.1. What are the SI units for mass, length, and volume?

1.2. What are the meanings and abbreviations for the prefixes kilo, centi, milli, micro, and nano?

1.3. Which SI unit and prefix would be most appropriate for measuring your body mass? Explain.

1.4. Which SI unit and prefix would be most appropriate to use with the diameter of a human hair? Explain.

1.5. What is the meaning and advantage of using scientific notation, e.g. 1.0×10^{-6}?

Information

Sometimes you need to convert from one unit to another. Engineers, health professionals, biologists, and other scientists find unit conversion to be a necessary part of their jobs. For example, the results of two measurements can only be compared directly if they have the same units.

For example, if you want to compare the price of gasoline in Europe (where gasoline is sold by the liter) and the United States (where gasoline is sold by the gallon), you need to convert liters to gallons, and in cooking it is helpful to be able to convert between cups and pints, pints and quarts, and teaspoons and tablespoons.

Also, when numerical values are calculated, the result must have the correct units. The units in the result are obtained by performing the arithmetic operations on the units as well as on the numbers. If the units obtained for the result are incorrect, then the value calculated also must be incorrect. Checking whether the units of the result are correct or not is a powerful method for validating a calculation. This validation is called *dimensional analysis* or *unit analysis*.

Unit conversion is accomplished by using equivalence statements to produce unit conversion factors. An *equivalence statement* is an equality that shows the relationship between two different units. A *conversion factor* is a ratio of units that equals 1. Since the conversion factor equals 1, it can multiply a quantity and change the units but not the actual physical magnitude of the quantity. Two conversion factors are obtained from an equivalence statement by dividing through by one side or the other.

For example, the equivalence statement for gallons and liters, Equation 1, produces two conversion factors, Equations 2 and 3, by dividing through by the quantity on one side or the other.

$$1 \text{ gallon} = 3.785 \text{ L} \tag{1}$$

$$\frac{1 \text{ gal}}{3.785 \text{ L}} = \frac{3.785 \text{ L}}{3.785 \text{ L}} = 1 = 0.2642 \text{ gal/L} \tag{2}$$

$$\frac{1 \text{ gal}}{1 \text{ gal}} = \frac{3.785 \text{ L}}{1 \text{ gal}} = 1 = 3.785 \text{ L/gal} \tag{3}$$

The conversions factors in Equations (2) and (3) both equal 1 because the numerator and denominator of each represent the same thing; they just have different units.

Suppose we purchase 5.00 gal of gasoline in the United States and 18.93 L of gasoline in France. To compare the amount of gasoline purchased in each situation, we need to express the amounts in common units. So convert liters to gallons by using the conversion factor in Equation (2) to show that 18.93 L and 5.00 gal are the same.

$$18.93 \text{ L } (0.2642 \text{ gal/L}) = 5.00 \text{ gal} \tag{4}$$

Exploration - 2

2.1. How many unit conversion factors result from a single equivalence statement? Explain.

2.2. (a) Multiply 15 gal by the conversion factor in Eq. 2 and do the arithmetic on the units as well as the numbers.

(b) Multiply 15 gal by the conversion factor in Eq. 3 and do the arithmetic on the units as well as on the numbers.

(c) Compare the results obtained in parts (a) and (b) and explain how doing the arithmetic on the units as well as the numbers identifies the correct conversion.

8

Application

1. The mass of a car is 1,365,000 g. Determine the mass in kg and in pounds. (1 kg = 2.205 lbs)
   (a) First estimate the answer without using a calculator, divide by 1000 to convert g to kg, then multiply by 2 to approximately convert kg to lbs. Write your estimate below. Making estimates is another way to validate your answers and inform you when you have made errors in the calculation.

   (b) Write the answer you obtained using a calculator.

2. A light pulse from a laser is 0.15 ns long. Determine the duration of this pulse in ps.
   (a) First identify whether a picosecond is longer or shorter than a nanosecond.

   (b) Based on your answer to (a) above identify whether the value in picoseconds will be larger or smaller than 0.15 ns.

   (c) Now calculate your answer and validate it by comparing with your answer to (b).

3. Express the volume 1.0 L in terms of cubic meters, m^3. Note that 1.0 L = 1000 cm^3, and the length conversion factor is 1 m/100 cm.
   (a) Noting that the volume of a cube 10 cm on a side is 1000 cm^3 (10 cm x 10cm x 10 cm), why do you need to apply the length conversion factor three times to produce the volume conversion factor 1 m^3/10^6 cm^3.

   (b) How many cubic meters are equivalent to 1.0L?

4.   Estimate the number of atoms that could comprise a baseball, given that the volume of a baseball is 200 cm³ and the volume of an atom is 0.004 nm³. Do the arithmetic on the units as well as the numbers to validate that you have made the correct unit conversions.

5.   You are the purchasing agent of a start-up biotechnology firm. If sucrose costs $11.80 per pound, and a bottle contains 5.00 kg, how much would you pay for a case of sucrose containing 12 bottles?

Got It!

1.   Use the equivalence statement (1 kg = 2.205 lbs) to check your response to Question 1 in the *What do you think?* section. Write the mass of both football players in kilograms. Who is heavier?

2.   Use the following equivalence statements to check your response to Question 2 in the *What do you think?* section. (1 m = 1.094 yd, 1 yd = 36 in) Write the height of both basketball players in meters. Who is taller?

3.   Obtain the appropriate equivalence statement from Tables 1 and 2 to check your response to Question 3 in the *What do you think?* section. Write the time for each skater in seconds. Who is faster?

02-1. The Nuclear Atom: What is the smallest particle of an element?

Atoms are the fundamental building blocks of all substances. Fig. 1 shows a representation of the atoms of three elements and gives the atomic symbols for them. Note that the nucleus and the surrounding electron cloud (e⁻) are not drawn to scale. The diameter of the electron cloud actually is 100,000 times larger than the diameter of the nucleus.

Exploration

1. What are the three particles that make up atoms?

carbon nucleus with 6 protons and 6 neutrons

6e⁻

6p 6n

$_{6}^{12}C$

carbon-12

2. Where are each type of component particles located in the atom?

silicon nucleus with 14 protons and 14 neutrons

14e⁻

14p 14n

$_{14}^{28}Si$

silicon-28

3. What information is provided by symbol, e.g. $_{26}^{56}Fe$?

4. What is the relationship between the number of electrons in any n⟨

iron nucleus with 26 protons and 30 neutrons

26e⁻

26p 30n

$_{26}^{56}Fe$

iron-56

Fig. 1 Atoms of carbon, silicon, and iron.

Information

All atoms of a given element have the same number of protons.

Atoms of the same element with different numbers of neutrons are called *isotopes*. Isotopes are distinguished by their different mass numbers, e.g. carbon-13 is an isotope of carbon-12.

Electrons can be removed from atoms producing an atomic ion that has a positive charge because it has more protons than electrons. A positively charged ion is called a *cation*. The charge of an ion is specified by a right superscript in the atomic symbol, e.g. Mg^{2+}.

Electrons also can be added to atoms producing a negatively charged ion. For example, O^{2-} has 2 more electrons than protons. A negatively charged ion is called an *anion*.

Got It!

1. Complete the entries in the Table I. The first row has been completed for you. In the table, Z = atomic number, which is the number of protons, A = mass number, which is the number of protons and neutrons, and N = number of neutrons. Use a Periodic Table to look up symbols, names, and atomic numbers where necessary.

Table I. Composition of Different Atoms

Name	Symbol	Z	A	N	Number of Electrons
Chlorine-37	$^{37}_{17}Cl$	17	37	20	17
Chloride-37	$^{37}_{17}Cl^-$				
	$^{35}_{17}Cl$				
Oxygen-16					
		6		6	
		6		7	
		11	23		10
Silver-107					46
Sulfur-32					18

02-2. The Mole and Molar Mass: Can you count atoms and molecules by weighing them?

Atoms combine to form molecules in chemical reactions. Keeping track of the number of atoms of each element involved in a chemical reaction is very important. This knowledge enables chemists to determine molecular formulas and mechanisms of reactions. It also determines how much of each reactant to use in a chemical reaction and how much product to expect.

What do you think?

1. Is it possible to determine the number of objects from their mass?

2. If so, explain how the number of objects can be determined from their mass. If not, explain why not.

Exploration: Solving a problem and applying the solution to chemistry.

You purchased 10 pounds of pennies in small pail at an estate sale for $10.00. That seemed like a good deal, pennies for a dollar a pound. Rather than tediously counting all the pennies, you decide to determine how many you have from the mass. You find that a single penny weighs 2.509 g and that 1 kg = 2.205 lb.

1. How can you calculate the number of pennies in the pail from the information that is given?

13

2. Transform you answer to Question 1 above to a mathematical equation that shows how to calculate the number of pennies in the pail. Use the following symbols:

N_p = the number of pennies,

M_t = the total mass of all the pennies, and

m_p = the mass of one penny.

3. How many pennies are in the pail that has exactly 10 lbs of pennies?

4. (a) What is the value of the pennies in the pail?

   (b) Explain whether you got a good deal or not since you purchased the pail of pennies for $10.00.

Information

Masses of atoms can be determined by a technique called mass spectrometry. Since the mass of an individual atom is very small, a special unit is used. This unit is called an *amu* (atomic mass unit) with 1 amu = 1.6605×10^{-24} g. The amu unit has this value in grams because it is defined as 1/12 the mass of a carbon-12 atom.

In dealing with naturally occurring samples, the average mass of all the isotopes is used because such samples contain all isotopes in their naturally occurring amounts.

5. Write an equation like the one you wrote for item 2 above that you can use to determine the number of atoms in a bar of platinum (Pt) given the mass of the bar and the average mass of a platinum atom.

6. How many atoms are there in a bar of platinum that has a mass of exactly 1 kg, given that the average mass of a platinum atom is 195.08 amu?

14

Information

In Question 6 above, you found that 1 kg of Pt contains 3.0871×10^{24} atoms. Dealing with such large numbers is not very convenient, so chemists invented a new unit for counting atoms. This unit is called a *mole*. This unit works just like the unit *dozen* for counting eggs. If you go to purchase a dozen eggs, you come back with 12 eggs. If you go to purchase a mole of platinum atoms, you come back with 6.0221×10^{23} platinum atoms. One mole of any substance contains 6.0221×10^{23} particles of that substance. The number of particles in a mole is so important that it is given a special name and symbol. It is called *Avogadro's number* and has the symbol N_A.

The mole is defined as the number of atoms in exactly 12 g of carbon-12. The mass of one mole of a substance is called the *molar mass*. The molar mass always is given in grams not kilograms.

Application

1. If you purchase a dozen apples, how many apples do you get?

2. If you purchase a mole of apples, how many apples do you get?

3. If a dozen apples costs $3.00, how much does a mole of apples cost?

4. If a dozen apples weighs 2.0 kg, how much does a mole of apples weigh?

5. Show how to determine the molar mass (in grams) of Pt given that the average mass of a platinum atom is 195.08 amu.

6. Show how to determine the molar mass (in grams) of Cl_2, given that the average mass of a chlorine molecule is 70.90 amu.

7. From your answers to 5 and 6, identify the relationship between the molar mass in grams and the average mass in amu of a particle comprising a substance. Note that the molar masses of all the elements are listed on the Periodic Table.

8. In Exploration Question 6, you found that 1 kg of Pt contains 3.0871×10^{24} atoms.
(a) Show how you can use Avogadro's number with this information to determine the number of moles of Pt atoms in 1 kg of Pt.

(b) Show how you can use the molar mass of Pt to determine the number of moles of Pt atoms in 1 kg of Pt. You found the molar mass of Pt in Application Question 5 above.

Got It!

If you have mastered this material you should be able to convert between any two of the following: mass, moles, and number of particles. To demonstrate your mastery, complete the following statements using the terms *molar mass* and *Avogadro's number*, use dimensional or unit analysis to demonstrate that your answer is correct, and then apply your understanding in solving the problem at the end.

1. To convert grams to moles, divide by _____
   as shown by the following unit analysis.

2. To convert moles to number of particles multiply by_____
   as shown by the following unit analysis.

3. To convert number of particles to moles divide by_____
   as shown by the following unit analysis.

4. To convert moles to grams, multiply by _____
   as shown by the following unit analysis.

5. You are the purchasing agent for a pharmaceutical company that manufactures cisplatin, which is a potent chemotherapy drug. The molecular formula for cisplatin is $PtCl_2(NH_3)_2$. You need to order platinum, Pt, and chlorine, Cl_2, to use in the production of cisplatin. Since platinum is expensive and chlorine is toxic, you do not want to purchase more of either one than is needed. From the molecular formula you know that to make 1 molecule of cisplatin, you need 1 platinum atom, and 2 chlorine atoms (i.e. 1 chlorine molecule).

(a) If you order 100 kg of platinum, how many moles of Pt will you receive?

(b) How many atoms of Pt will you receive?

(c) How many moles of chlorine, Cl_2, are needed to react with all the platinum?

(d) How many kg of chlorine are needed to go with the 100 kg of platinum?

(e) How many moles of cisplatin can be made from these amounts of platinum and chlorine?

03-1. Naming Compounds: What's in a name?

Atoms combine to form molecules, for example, $O + O \rightarrow O_2$ and $C + O \rightarrow CO$. O_2 is a molecule of the element oxygen, and CO is a molecule of the compound carbon monoxide. A *compound* is formed from two or more different elements.

In order to talk about compounds in a meaningful way, they need to have names that tell us something about their composition. The composition of a compound is represented by its molecular or chemical formula. In a molecular or chemical formula, the elements forming the compound are designated by the symbol for the element, and the number of atoms of that element in a molecule of the compound is given by a subscript. For example, CO_2, represents a molecule of the compound carbon dioxide that is composed of 1 carbon atom and 2 oxygen atoms.

What do you think?

Almost everyone knows that water is H_2O. Many waiters in restaurants write H_2O when you order water. Water is the common name for H_2O. Invent a name for water that reveals that it is composed of two hydrogen atoms and one oxygen atom.

Exploration – 1: Names of Binary Covalent Compounds

In a covalent compound atoms are bound together by sharing one or more pairs of electrons.

Table I. Names of Some Binary Covalent Compounds

Molecular Formula	Name
HF	Hydrogen fluoride
HCl	Hydrogen chloride
H_2S	Hydrogen sulfide
CO	Carbon monoxide
CO_2	Carbon dioxide
CCl_4	Carbon tetrachloride
N_2O	Dinitrogen oxide
N_2O_5	Dinitrogen pentoxide
PBr_3	Phosphorous tribromide
PBr_5	Phosphorous pentabromide
SF_6	Sulfur hexafluoride

1.1. As evident from Table I, how many different elements are there in a binary compound?

1.2. Which element is named first, the one that is less electronegative or the one that is more electronegative? The more electronegative elements are to the right on the Periodic Table. *Electronegativity* refers to the attraction an atom has for electrons.

1.3. What ending (suffix) is applied to the root of the more electronegative element in the second part of the name?

1.4. What prefixes are used to indicate, when necessary, that the number of atoms is 1, 2, 3, 4, 5, or 6?

Information

Some elements always combine with others in certain ratios so it is not necessary to specify explicitly the number atoms of each element in the compound. For example, hydrogen always combines with the halogens (Group VIIA) in a 1:1 ratio, and with the chalcogenides (Group VIA) in a 2:1 ratio. So H_2S is called hydrogen sulfide, and water is called hydrogen oxide, not *dihydrogen oxide*.

Not all compounds are covalent. Some are ionic. In an ionic compound, one or more electrons are transferred from one atom to another to form ions, see Figure 1. This transfer occurs because one atom is much more electronegative than the other.

The atom that lost the electron(s) has a positive charge and is called a *cation*. The atom that gained electron(s) has negative charge as a result and is called an *anion*.

Fig. 1. Formation of an ionic compound NaCl.

The two ions are held together by the electrostatic attraction between the positive and negative charges.

Cations are formed from metals that are not very electronegative on the left side of the Periodic Table. Anions are formed from the electronegative nonmetals on the right side of the Periodic Table.

Exploration – 2: Names of Binary Ionic Compounds

Table II. Names of Some Binary Ionic Compounds

Chemical Formula	Name	Cation Charge	Anion Charge
NaCl	Sodium chloride		
KBr	Potassium bromide		
$MgCl_2$	Magnesium chloride		
BaI_2	Barium iodide		
CaO	Calcium oxide		
CuCl	Copper(I) chloride		
$CuCl_2$	Copper(II) chloride		
$CrCl_3$	Chromium(III) chloride		
Al_2O_3	Aluminum oxide		
Fe_2O_3	Iron(III) oxide		
FeO	Iron(II) oxide		

2.1. According to Table II, which element is named first in an ionic compound, the metal or the nonmetal?

2.2. What ending (suffix) is applied to the root of the nonmetal in the second part of the name?

2.3. When a metal ion can form more than one kind of cation, how is the charge on the ion indicated?

2.4. What are some similarities and differences in the names of binary covalent compounds and binary ionic compounds?

21

Information

Group IA and Group IIA elements always give up 1 and 2 electrons, respectively.

When forming ionic compounds, oxygen and other Group VIA elements always take 2 electrons, and the halogens in Group VIIA always take 1 electron.

Some transition metals can lose varying numbers of electrons to form cations with different charges. The possibilities you should know are Cr (+2 and +3), Fe (+2 and +3), Co (+2 and +3), and Cu (+1 and +2). In a compound's name, the charge on the metal is indicated by the Roman numerals in parentheses following the name of the metal.

Got It!

If you have mastered this material, you should be able to complete the following.

1. Write the charges on the cations and anions in the appropriate columns of Table II.

2. Write the formulas for the following compounds.
   (a) sulfur dioxide

   (b) silicon tetrachloride

   (c) chromium(III) oxide

3. Write the names of the following compounds.
   (a) $CoCl_2$

   (b) $CoCl_3$

   (c) K_2S

   (d) SO_3

4. Explain why it is not necessary in Table II to name
   (a) $MgCl_2$ magnesium dichloride.

   (b) $CuCl_2$ copper(II) dichloride

22

03-2. Determining Empirical and Molecular Formulas: Why can different molecules have the same percent composition by mass?

Chemical analysis reveals the composition of a compound in terms of the mass percent of each element present. The empirical formula can be deduced form this information, and when combined with other information, the molecular formula of the compound can be determined.

Exploration

Your challenge is to complete Table I below and identify what the pairs of molecules listed in the table have in common. You should be able to do this right after you have written the molecular formulas, but also calculate the percent composition by mass as well. Write the percent composition using 3 significant figures. Use the first row to check that your procedure is correct since this row for acetylene has been done for you.

To complete this task, you need to be able to convert structural formulas into molecular formulas. You also need to be able to determine the molar mass of compounds, and the percent by mass of each element in 1 mole of the compound

Table I. Compounds with Something in Common

Name	Structural Formula	Molecular Formula	Percent by Mass Carbon	Hydrogen
Pair 1				
Acetylene	$H-C{\equiv}C-H$	C_2H_2	92.2%	7.76%
Benzene				
Pair 2				
Ethene				
Cyclohexane				

1. What are the common features of the four compounds in Table I that you can identify from the table?

2. How are the molecular formulas of these compounds similar and how are they different?

3. If a chemical analysis provides you with the mass percent composition of an unknown compound, could you determine the molecular formula of the compound from this information? Explain why or why not based on the information in Table I.

4. Why do the two compounds in Pair 1 and the two compounds in Pair 2 have the same mass percent composition?

5. The formula that expresses the correct ratio of carbon atoms to hydrogen atoms for both compounds in Pair 1 is CH. What is the corresponding formula that expresses the correct ratio of carbon atoms to hydrogen atoms for both compounds in Pair 2?

Information

The formulas in Question 5 above, CH and CH$_2$, are called empirical formulas. An *empirical formula* for a compound gives the symbols for the elements in the compound, and the subscripts provide the ratio of the atoms of each element using the smallest possible whole numbers (integers). The *molecular formula* is some multiple of the empirical formula. The empirical formula can be determined from the mass percent composition, but additional information, for example, the molar mass, is needed to determine the molecular formula from the empirical formula.

Application

Phosphorous forms two compounds with oxygen. One contains 56.34% phosphorous and 43.66% oxygen. Use this information to determine its empirical formula.

To obtain the empirical formula, we need the ratio of O atoms to P atoms in the compound. The mass percent composition tells us that if we had 100 g of the compound, 56.34 g would be P and 43.66 g would be O. We can use this information to get the ratio that we need.

1. How many moles of P atoms are there in 56.34 g of P? Show that the answer is 1.819 mol P.

2. How many moles of O atoms are there in 43.66 g of O? Show that the answer is 2.729 mol O.

3. What is the smallest whole number ratio of O atoms to P atoms? From Items 1 and 2 we see that the ratio is
$$\frac{\text{O atoms}}{\text{P atoms}} = \frac{2.729}{1.819}.$$
We need to convert these numbers into whole numbers without changing the value of the ratio, so divide both the numerator and the denominator by the smallest,
$$\frac{\text{O atoms}}{\text{P atoms}} = \frac{2.729/1.819}{1.819/1.819} = \frac{1.500}{1.000},$$

and then multiply both the numerator and denominator by 2.

$$\frac{\text{O atoms}}{\text{P atoms}} = \frac{1.500}{1.000} = \frac{2 \times 1.500}{2 \times 1.000} = \frac{3}{2}$$

4. What is the empirical formula for this compound?

5. Show that the molar mass associated with the empirical formula is 109.95 g/mol.

6. The molar mass of this compound was determined to be 219.9 g/mol. How many multiples of the empirical formula does it take to produce a mass of 219.9 g/mol?

7. What is the molecular formula of this compound?

Got It!

Another oxide of phosphorous contains 43.64% phosphorous and 56.36% oxygen. It has a molar mass of 283.89 g/mol. Determine its empirical and molecular formulas.

04-1. Reaction Stoichiometry: How much do I need to get what I want?

Determining the amounts of reactants and products in a chemical reaction is called *reaction stoichiometry*. *Stoichiometry* is a word that comes from the Greek and means *measuring the pieces*. Balanced reaction equations are used to relate the amounts of products produced by various amounts of reactants.

You have a stoichiometry problem whenever you are asked, *How much is produced? How much is used? How much is needed? What is the reaction yield? What is the limiting reactant?* To answer such questions, you need to determine the quantity of one reactant or product from the quantity of another reactant or product.

There are three keys to stoichiometry problems. You need to be able to (1) balance the reaction equation, (2) recognize that the amounts in moles of products and reactants are in the same proportions as the stoichiometric coefficients in the balanced equation, and (3) make various conversions. *What are these conversions?* you ask. They are summarized in Table I. Don't memorize this table rather use unit analysis to figure out the conversion.

Table I. Conversions Found in Stoichiometry Problems

Quantities	Conversion Factor	Unit Analysis
grams to moles	Divide by molar mass	g / (g /mol) = mol
moles to grams	Multiply by molar mass	mol x (g/mol) = g
gas volume to moles	Multiply by P/(RT).	L x (mol/L) = mol
moles to gas volume	Multiply by RT/P.	mol x (L/mol) = L
liquid volume to mass	Multiply by the density.	mL x (g/mL) = g
solution volume to moles of solute	Multiply by the molar concentration.	L x (mol/L) = mol

A specific strategy, like following the steps below, can be very helpful in solving stoichiometry problems.

Step 1: Write the balanced reaction equation if it is not given to you.

Step 2: Write the amounts you are given and identify what you need to find immediately below the reaction equation. Include units!

Step 3: Convert quantities to moles and write the values below the reaction equation as well.

Step 4: Set up a mole ratio: the quantity of the compound that you need to find divided by the quantity of the compound that is given. Set up another ratio involving the stoichiometric coefficients of these two compounds in the balanced reaction equation. These two ratios must be equal because the moles of reactants and products must be in proportion to the stoichiometric coefficients.

Step 5: Solve for what you need to find in the ratio, and convert moles to the units requested.

Example

Climate change, partially caused by the production of greenhouse gases like carbon dioxide in the combustion of fossil fuels, is a major contemporary issue. To understand the arguments on both sides of this issue, one needs to be able to determine the amount of carbon dioxide produced by different fossil fuels. Follow the procedure below to determine the amount (in grams) of carbon dioxide produce by burning 1.00 kg of propane.

Step 1	$C_3H_8(g) + 5O_2(g) \rightarrow 3CO_2(g) + 4H_2O(g)$
Step 2	1000 g ?
Step 3	1000 g / (44.1 g/mol) = 22.7 mol C_3H_8
Step 4	$\dfrac{3\,mol\,CO_2}{1\,mol\,C_3H_8} = \dfrac{x}{22.7\,mol\,C_3H_8}$
Step 5	x = 68.1 mol CO_2 → 68.1 mol x 44.0 g/mol = 3000 g CO_2

Exploration

1. Examine the five step strategy for solving stoichiometry problems that is followed in the above example.

   (a) Why is writing the balanced reaction equation an important part of solving stoichiometry problems?

   (b) What purpose is served by Steps 2 and 3?

   (c) In Step 4, why must the mole ratio of the amounts equal the ratio of the stoichiometric coefficients?

2. Are there any steps in the strategy that you feel are not useful? Explain.

3. Can you improve the strategy by changing the steps or changing the order of the steps? Explain.

4. What are 5 insights about solving stoichiometry problems that you gained by reading the introductory material and examining the strategy and its use in the example?

Application

1. When 0.5 mol of ammonium nitrate decomposes, how many moles of water are produced?

$$\underline{}NH_4NO_3(s) \rightarrow \underline{}N_2(g) + \underline{}H_2O(g) + \underline{}O_2(g)$$

2. How many moles of ammonium nitrate must decompose to produce 96 g of oxygen?

3. How many grams of a Freon, CCl_2F_2, are produced by reacting 231 g of carbon tetrachloride, CCl_4, with hydrofluoric acid, HF?

Got It!

1. Cisplatin, $PtCl_2(NH_3)_2$, is a drug used in the treatment of cancer. It can be synthesized by the reaction of K_2PtCl_4 with ammonia. What mass of ammonia in grams is needed to produce 1.00 kg of cisplatin?

2. Nitroglycerin is unstable and decomposes with explosive violence because it releases much energy and forms a large amount of gaseous products. How many moles of gas will be produced by 227 g of nitroglycerin exploding?

$$4\ C_3H_5(NO_3)_3(l) \rightarrow 12\ CO_2(g) + 10\ H_2O(g) + 6\ N_2(g) + O_2(g)$$

Additional Practice Problems

1. A candy bar contains 43.4 g of sucrose (sugar, $C_{12}H_{22}O_{11}$). When the candy bar is eaten, the sucrose is metabolized to carbon dioxide and water.
   (a) Write the balanced reaction equation for sucrose metabolism.

   (b) Determine the moles of sucrose in the candy bar and the moles of carbon dioxide and water that are produced.

(c) Determine the masses of carbon dioxide and water that are produced.

2. Sulfur dioxide gas is produced by coal-fired power plants. It combines with water in the air to produce sulfurous acid, which is a component in acid rain. Sulfur dioxide can be removed from smokestacks by treatment with limestone and oxygen, according to the following unbalanced reaction equation.

$$\_\_SO_2(g) + \_\_CaCO_3(s) + \_\_O_2(g) \rightarrow \_\_CaSO_4(s) + \_\_CO_2(g)$$

(a) What mass (kg) of calcium carbonate (limestone) is required to remove 1.0 kg of sulfur dioxide?

(b) What mass (kg) of calcium sulfate is produced in removing 1.0 kg of sulfur dioxide?

(c) What mass (kg) of carbon dioxide is produced in removing 1.0 kg of sulfur dioxide?

3. The following reaction is used to produce iron from iron ore. How may metric tons of iron can be produced from 1.0 metric ton of iron ore? A metric ton is exactly 1000 kg.

$$\_\_Fe_2O_3(s) + \_\_CO(g) \rightarrow \_\_Fe(s) + \_\_CO_2$$

04-2. Limiting Reactants: How much will I get with what I have?

Reactants are not always present in the exact amounts indicated by a balanced reaction equation. Usually in laboratory reactions or industrial processes, one reactant is more expensive or less readily available than others. The cheaper or more available reactants are used in excess to ensure that the more expensive or rarer substance is completely converted to product.

The limiting reactant is the reactant that is completely converted to product. Once the limiting reactant has been used up, no more product can form. Given information about the amounts of reactants present, you need to be able to determine which one is the limiting reactant. The moles of product formed are determined by the number of moles of the limiting reactant present at the beginning of the reaction.

As demonstrated below, an analogy for the limiting reactant can be made with the ingredients of a cheese sandwich.

What do you think?

In making cheese sandwiches or other food dishes, there often is a limiting ingredient. Suppose you have 40 bread slices, 25 cheese slices, and 30 lettuce leaves, how many cheese sandwiches could you make? Each sandwich must have 2 slices of bread, 1 slice of cheese, and 1 leaf of lettuce. This recipe can be written like a reaction equation.

2 bread slices + 1 cheese slice + 1 lettuce leaf → 1 sandwich

33

Exploration

1. Which ingredients were left over after you made the cheese sandwiches?

2. Which ingredient was the limiting ingredient?

3. What procedure did you use to identify the limiting ingredient?

4. Try your procedure and see if it works with molecules. Molecules are just like cheese sandwiches. Suppose you have 4 nitrogen molecules and 6 hydrogen molecules. How many ammonia molecules can you make?

$$N_2 + 3H_2 \rightarrow 2NH_3$$

Fig. 1. 4 N_2 molecules and 6 H_2 molecules

5. How can you use the stoichiometric coefficients in the balanced reaction equation to help you identify the limiting reactant?

Information

One way to identify the limiting ingredient or limiting reactant is to see how much product can be made by each. For example, 40 bread slices can make 20 sandwiches, 25 cheese slices can make 25 sandwiches, and 30 lettuce leaves can make 30 sandwiches. The bread slices make the fewest number of sandwiches so they are the limiting ingredient.

Similarly 8 ammonia molecules can be made from 4 nitrogen molecules, and 4 ammonia molecules can be made from 6 hydrogen molecules, so in Question 5, hydrogen is the limiting reactant because it produces the fewest ammonia molecules.

Application

1. Sketch a diagram in the empty space by Exploration Question 4 to show the ammonia molecules and the left over nitrogen molecules. Use circles shaded differently to represent the N and H atoms.

2. A rocket fuel, hydrazine N_2H_4, is produced by the reaction of chlorine, Cl_2, with sodium hydroxide, NaOH, and ammonia, NH_3. How many moles of hydrazine can be produced using excess sodium hydroxide, 15 mol chlorine, and 15 mol ammonia?

$$2NaOH + Cl_2 + 2NH_3 \rightarrow N_2H_4 + 2NaCl + 2H_2O$$

Got It!

Dialuminum hexachloride, Al_2Cl_6, is used in many industrial processes. It is made by treating scrap aluminum metal (Al) with chlorine gas (Cl_2).

1. Write the balanced reaction equation for the production of dialuminum hexachloride.

2. What is the limiting reactant if 710 g of chlorine gas reacts with 270 g of aluminum?

3. How much dialuminum hexachloride is produced by this reaction?

4. How much of each reactant is left over when the reaction is complete?

5. If only 750 g of dialuminum hexachloride was produced, what was the percent yield of the reaction? Note: percent yield = (actual amount produced / maximum amount that could be produced) times 100.

Additional Practice Problems

1. Aspirin is produced by the reaction of salicylic acid and acetic anhydride. If 100 g of each of these reactants are used, what is the maximum amount of aspirin (in g) that could be produced?

$$2C_7H_6O_3(s) + C_4H_6O_3(l) \rightarrow 2C_9H_8O_4(s) + H_2O(l)$$

2. The reaction between lithium hydroxide and carbon dioxide has been used to scrub carbon dioxide from spacecraft atmospheres. The products are lithium carbonate and water. Identify the excess reactant and the amount that remains when 10 g of lithium hydroxide and 20 g of carbon dioxide react.

3. Suppose methanol, a clean burning fuel, could be produce by reacting coal (mostly carbon) with water. What volume of methanol could be produced by reacting 1.0 kg of carbon with 1.0 kg water? The density of liquid methanol is 0.79 g/mL.

$$2C(s) + 2H_2O(g) \rightarrow CH_3OH(l) + CO(g)$$

4. Farmers use urea as a nitrogen fertilizer. In the presence of water it decomposes to ammonia, which is readily converted into nitrates that plants can utilize.

$$CO(NH_2)_2 + H_2O \rightarrow 2NH_3 + CO_2$$

If a urea fertilizer were applied at the rate of 20 lbs/acre just prior to a rainfall of 0.25 inches, how much of the fertilizer per acre would be converted to ammonia, and how much would remain, assuming all the water reacted with the urea? A rainfall of 0.25 inches means that any impermeable flat area would be covered by water to a depth of 0.25 inches.

(1 acre = 4047 m^2, 1 in = 2.54 cm, 1 lb = 454 g)

05-1. Types of Chemical Reactions: What happens in a chemical reaction?

What do you think?

Describe what you think happens in a chemical reaction, and identify three different types of chemical reactions that you have encountered.

Information

Chemical compounds can react with each other and form new compounds. You need to visualize or represent these reactions in three ways: macroscopic, nanoscopic, and symbolic. The macroscopic representation is what you see with your eyes. The nanoscopic representation is what the atoms and molecules are doing, and the symbolic representation is a chemical reaction equation. Most likely you can find pictures and diagrams in your textbook that show what chemical reactions look like at the macroscopic and nanoscopic levels. This activity is directed at helping you understand reaction equations and identify some different types of chemical reactions.

Exploration

1. Some different types of chemical reactions are listed Table I. In order to make sense of the information in this table, you need to understand the symbols and words that are used. Use your textbook to provide an explanation for each of the following:

   (a) the symbols (s), (g), (l), and (aq).

(b) an ionic compound.

(c) to dissolve.

(d) a solution.

(e) an acid and a base

(f) hydronium ion and hydroxide ion.

Table I. Types of Reactions

Reaction Types / Characteristics	Examples
Dissociation *an ionic compound dissolves in water to produce ions*	$Ca(NO_3)_2(s) \rightarrow Ca^{2+}(aq) + 2NO_3^-(aq)$ *calcium nitrate dissociates*
Precipitation *a solid forms when solutions are mixed*	$AgNO_3(aq) + KCl(aq) \rightarrow AgCl(s) + KNO_3(aq)$ *silver nitrate reacts with potassium chloride*
Acid Ionization *an acid reacts with water to produce hydronium ions*	$HCl(g) + H_2O(l) \rightarrow H_3O^+(aq) + Cl^-(aq)$ *hydrochloric acid ionizes*
Base Ionization *a base reacts with water to produce hydroxide ions*	$NH_3(g) + H_2O(l) \rightarrow NH_4^+(aq) + OH^-(aq)$ *ammonia ionizes*
Base Dissociation *a base dissolves in water to produce hydroxide ions*	$NaOH(s) \rightarrow Na^+(aq) + OH^-(aq)$ *sodium hydroxide dissociates*
Acid – Base Neutralization *an acid and a base react to produce water and a salt*	$HCl(aq) + NaOH(aq) \rightarrow H_2O(l) + NaCl(aq)$ *hydrochloric acid neutralizes sodium hydroxide*
Oxidation – Reduction *electrons are transferred from one reactant to another*	$2Ag^+(aq) + Cu(s) \rightarrow 2Ag(s) + Cu^{2+}(aq)$ *electrons are transferred from copper atoms (oxidation) to silver ions (reduction)*

2. What are the characteristics of the reactions in Table I that you can use to identify the reaction? Identify the nature of the reaction and the general types of substances that react and are produced for each of the following reactions. You need to remember these characteristics, and writing them without looking at Table I will help you remember.

(a) dissociation

(b) precipitation

(c) acid ionization

(d) base ionization

(e) base dissociation

(f) acid-base neutralization

(g) oxidation – reduction

3. Which three types of reactions in Table I involve the transfer of a hydrogen ion, H⁺, from one reactant to another?

4. Explain how the oxidation – reduction equation in Table I can be balanced when the charge on the silver ion is +1 and the charge on the copper ion is +2.

5. In the precipitation example in Table I, the reactants are ionic compounds so they dissociate in water. Consequently KCl(aq) means K⁺(aq) + Cl⁻(aq) and AgNO₃(aq) means Ag⁺(aq) + NO₃⁻ (aq). Write the reaction equation using only those ions that produce the precipitate AgCl(s).

Information

Notice that all the reaction equations in Table I are balanced. *Balanced* means that the number of atoms of each element and the number of electrons are the same on both the reactant side and the product side of the equation. Reaction equations must be balanced because electrons and atoms are not created or destroyed in a chemical reaction.

The overall reaction equation is written for the precipitation example in Table I. All the compounds in this example are ionic so in aqueous solution they all dissociate except for silver chloride, AgCl, which is not soluble in water. Consequently KCl(aq) means K⁺(aq) + Cl⁻(aq) and AgNO₃(aq) means Ag⁺(aq) + NO₃⁻(aq). The precipitation reaction involves only the silver ions and the chloride ions. In the reaction equation that is written below, the potassium ions and the nitrate ions are not included because they are not involved in the precipitation reaction. The ions that are not involved are called *spectator ions*. A reaction equation that includes only those ions that are involved is called the *net ionic equation*. The net ionic equation, which should be your answer to Question 5 above, for the precipitation of silver chloride is

$$Ag^+(aq) + Cl^-(aq) \rightarrow AgCl(s).$$

As with precipitation reactions, net ionic equations can be written for acid-base neutralization reactions. For the example given in Table I, hydrochloric acid ionizes and sodium hydroxide dissociates in aqueous solution, so the net ionic equation is

$$H_3O^+(aq) + OH^-(aq) \rightarrow 2H_2O(l),$$

and the sodium and chloride ions are spectator ions.

Application

Complete and balance the following reactions, and identify the reaction type.

1. $Na_2SO_4(s) \rightarrow \quad Na^+(aq) \quad + \quad SO_4{}^{2-}(aq)$

2. $CO_3{}^{2-}(aq) \quad + \quad H_2O(l) \quad \rightarrow \quad H_2CO_3(aq) \quad + \quad OH^-(aq)$

3. $Fe^{3+}(aq) \quad + \quad OH^-(aq) \quad \rightarrow \quad Fe(OH)_3(s)$

4. $H_3PO_4(aq) \quad + \quad NaOH(s) \quad \rightarrow \quad Na_3PO_4(aq) \quad + \quad _____$

5. $HOCl(aq) \quad + \quad H_2O(l) \quad \rightarrow \quad _____ \quad + \quad ClO^-(aq)$

6. $Ca(s) \quad + \quad O_2(g) \quad \rightarrow \quad CaO(s)$

7. $Sr(OH)_2(aq) \quad \rightarrow \quad Sr^{2+}(aq) \quad + \quad _____$

Got It!

1. Lead(II) nitrate and hydrochloric acid react to produce a lead chloride precipitate.

   (a) Write the dissociation equation for lead(II) nitrate.

   (b) Write the acid ionization equation for hydrochloric acid.

   (c) Write the net ionic equation for the precipitation of lead(II) chloride.

2. Write the dissociation equation for sodium carbonate, Na_2CO_3, and the net ionic equation for the reaction of the carbonate ion, which is a base, with water.

3. A zinc metal rod, Zn, immersed in a solution of copper sulfate, $CuSO_4(aq)$, will react producing Zn^{2+} ions in solution and solid copper metal on the rod. What type of reaction is this? Write the net ionic equation for it.

4. Write the equations that describe what happens (a) when nitric acid is added to water, (b) when sodium hydroxide is added to water, and (c) when the two solutions are mixed to neutralize each other.

44

05-2. Solutions: What are they, what do they look like, and how are they prepared?

What do you think?

Identify three solutions that you have encountered that involve water as the solvent.

Exploration 1: Electrolytes and Nonelectrolytes

Electrolyte   Nonelectrolyte

1.1. The above diagram represents sodium chloride, NaCl, which is an ionic compound, and methanol, CH_3OH, which is a covalent compound, dissolving in water. Look carefully and identify the sodium and chloride ions, methanol molecules, and water molecules in this diagram. Which compound is an *electrolyte* and which one is a *nonelectrolyte*? Identify one similarity and one difference between these two types of solutions.

1.2. Which part of the water molecule is closest to the sodium ions in solution? Explain why.

1.3. Which part of the water molecule is closest to the chloride ions in solution? Explain why.

Information

Many of the materials that you encounter are solutions. Some examples of solutions are air, blood, beverages, and metal alloys like brass. A *solution* is a homogeneous mixture of two or more substances. The *solvent* is the component of the solution that is present in the largest amount. An *aqueous solution* is one that has water as the solvent. The *solute* is the substance, which is present in a smaller amount, dissolved in the solvent. There may be more than one solute in a solution. The *concentration* of a solution specifies the amount of solute present in some amount of solution or solvent.

Solutions can be solid, liquid, or gas. Examples of these are steel, gasoline, and air, respectively. The solute dissolved in a liquid solution initially can be a solid like sodium chloride, a liquid like ethanol, or a gas like oxygen. These dissolve in water to make salt water, an alcoholic beverage, or oxygenated water that supports fish life.

Substances that dissolve in water can be classified according to their ability to conduct electricity when in solution. These substances are *strong electrolytes* if they conduct electricity well, *weak electrolytes* if they conduct electricity poorly, and *nonelectrolytes* if they do not conduct electricity.

Many ionic compounds are strong electrolytes. When an ionic compound like sodium chloride, $NaCl$, dissolves in water, the cation and anion separate, and the motion of these charged particles through the solution serves to conduct electricity.

Covalent compounds like methanol or sugar are nonelectrolytes because when they dissolve in water they do not separate into charged particles. All the atoms remain covalently bonded to each other.

46

Some compounds are in between these two extremes of ionizing completely or not at all. Only a fraction of the molecules ionize, and only a few ions are produced that serve to conduct a small amount of electricity for these weak electrolytes.

Exploration 2: Preparing Solutions of Known Concentration

2.1. The picture to the right shows that 11.1 g of calcium chloride, $CaCl_2$ are being added to a volumetric flask to make 250.0 mL of solution. How many moles of calcium chloride is 11.1g?

2.2. If 11.1 g of calcium chloride are used to make 250.0 mL of solution,
(a) what is the concentration of this solution in moles/liter of solution?

(b) How many moles of calcium ion would there be in one liter of a calcium chloride solution with this concentration?

(c) How many moles of chloride ion would there be in one liter of a calcium chloride solution with this concentration?

Information

Solution concentration is commonly expressed as moles of solute in one liter of solution. This concentration is called the *molarity* of the solution and is represented by the letter M or by square brackets [ ] around a molecular formula. M stands for moles per liter. For example, [NaOH] = 1.5 M, so 1.5 M is read as 1.5 molar, which means that there are 1.5 moles per liter of solution. The molarity of a solution is calculated using the following formula, where n_{sol} = moles of solute and V_{soln} = volume of solution.

$$M = n_{solute} / V_{solution} \qquad (1)$$

2.3. The picture to the right shows that a 100.0 mL volumetric flask has been filled with 0.100 M potassium dichromate solution, $K_2Cr_2O_7$. This solution is to be transferred to the large 1.00 L volumetric flask, which then will be filled with water to the 1.00 L mark, producing 1.00 L of potassium dichromate solution.

(a) How many moles of potassium dichromate are in 100.0 ml of 0.100 M solution?

(b) How many moles of potassium dichromate will end up in the 1.00 L volumetric flask?

(c) What will the molar concentration of the potassium dichromate solution be in the 1.00 L flask after water has been added to the 1.00 L mark?

(d) What will the concentration of potassium ions be?

Information

In preparing solutions by dilution of a more concentrated solution, the moles of solute present in the concentrated solution is n, where $n = M_cV_c$ as seen by rearranging Equation (1). M_c is the molarity of the concentrated solution, and V_c is the volume of the concentrated solution. These moles of solute end up in the diluted solution, so n also must equal M_dV_d, where M_d is the molarity of the diluted solution, and V_d is the volume of the diluted solution. This conclusion can be represented by Equation (2). Given any three of the quantities in Equation (2), the remaining one can be calculated.

$$M_cV_c = M_dV_d \qquad (2)$$

Exploration 3: Relating Amounts of Reactants and Products in Solution

How many mL of a 3.0 M hydrochloric acid solution are needed to exactly neutralize 250.0 mL of 6.0 M sodium hydroxide solution?

3.1. Write the balanced reaction equation for this acid base reaction.

3.2. How many moles of sodium hydroxide are there in 250.0 mL of the 6.0 M solution?

3.3. How many moles of hydrochloric acid are needed to exactly neutralize the amount of sodium hydroxide you reported in Question 3.2?

3.4. What volume of the 3.0 M hydrochloric acid solution will contain the number of moles of hydrochloric acid that you reported in Question 3.3?

Information

You can solve solution stoichiometry problems like the one in Exploration 3 in the same way that you solve other stoichiometry problems. The following five step strategy is recommended.

(1) Write the balanced reaction equation.

(2) Write the amounts you are given, and what you need to find below each compound in the reaction equation.

(3) Convert the amounts to moles. In solution problems, you generally need to multiply volume in liters times molar concentration.

(4) Use the ratio of stoichiometric coefficients to find what you need to find.

(5) Convert moles to the units requested. In solution problems, if volume is requested, you divide the number of moles by the molar concentration, as you can see by using unit analysis.

Got It!

1. Identify one of the following as a strong electrolyte, one as a weak electrolyte, and one as a nonelectrolyte.

   propanol, $CH_3CH_2CH_2OH$

   acetic acid, CH_3COOH

   nitric acid, HNO_3

2. Identify the ions present in a solution of 0.12 M barium chloride, $BaCl_2$, and their concentrations.

3. Calculate the molarity of a 250.0 mL solution made with 0.25 moles of ammonium nitrate, a strong electrolyte.

4. Calculate the concentration of a solution produced by diluting 10.0 mL of 12.5 M hydrochloric acid to 500.0 mL.

5. Determine how much 12.0 M HCl you need to use to produce 250.0 mL of a 3.00 M solution.

6. What volume of a 2.06 M potassium permanganate, $KMnO_4$, solution contains 322 g of solute?

7. What mass in grams of barium hydroxide, $Ba(OH)_2$, is required to neutralize completely 25.0 mL of 0.125 M nitric acid, HNO_3?

Additional Practice Problems

1. A flask contains 0.25 M ammonium sulfate, $(NH_4)_2SO_4$. What ions are present in solution, and what are their concentrations?

2. What is the mass in grams of the solute in 250.0 mL of a 0.0125 M solution of potassium permanganate, $KMnO_4$?

3. You need to make a 0.500 M solution of nickel(II) sulfate, $NiSO_4$. How many grams of the hydrate, $NiSO_4 \cdot 6H_2O$, do you need to put into a 500.0 mL volumetric flask?

4. If 5.00 mL of a 0.0250 M solution of sodium phosphate, Na_3PO_4, is diluted to 100.0 mL with pure water, what is the concentration of the sodium ions in the resulting solution?

5. How many grams of baking soda, sodium hydrogen carbonate, are needed to neutralize 125 mL of 6 M hydrochloric acid that spilled on the lab bench?

$$NaHCO_3(s) + HCl(aq) \rightarrow NaCl(aq) + CO_2(g) + H_2O(l)$$

6. How many grams of baking soda, see the previous question, are needed to completely neutralize 125 mL of 6 M sulfuric acid, H_2SO_4, that spilled on the lab bench?

7. Vitamin C, $C_6H_8O_6$, can be oxidized by bromine, Br_2.

$$C_6H_8O_6(aq) + Br_2(aq) \rightarrow C_6H_6O_6(aq) + 2HBr(aq)$$

It took 27.85 mL of 0.102 M Br_2 solution to react completely with a 1.00 g tablet of Vitamin C with added filler.
(a) What is the mass in grams of vitamin C in the tablet?

(b) What percent of the tablet is vitamin C?

8. A fruit juice drink contains an unknown amount of citric acid. It is found that the citric acid in 10.0 mL of the drink can be neutralized by 7.55 mL of 0.125 M sodium hydroxide. What amount of citric acid, in grams, is there in 1.00 mL of the drink? The neutralization reaction is

$$C_3H_5O(COOH)_3(aq) + 3NaOH(aq) \rightarrow Na_3C_3H_5O(COO)_3(aq) + 3H_2O(l).$$

06-1. Energy: How do you know how much you have gained or lost?

Richard Feynman, a Nobel laureate in physics, said, "energy is a mysterious thing; ...we have no knowledge of what energy is." [R.P.Feynman, R.B. Leighton, and M. Sands, *The Feynman Lectures on Physics* (Addison-Wesley, 1963) pp. 4-1 to 4-2]

Einstein recognized that mass, m, and energy, E, are related by the speed of light, c. He proposed that

$$E = mc^2. \tag{1}$$

This equation can be transformed to $m = E/c^2$ and interpreted to mean that mass is just another form of energy. In fact the *Big Bang Theory* postulates that the universe was created by a sudden burst of energy, followed by expansion and cooling that produced elementary particles with mass, and subsequently protons, neutrons, atoms, and molecules. From this perspective, energy is one of the fundamental components that make up the universe, along with elementary particles like quarks.

The concept of energy is useful because it is a numerical quantity that is conserved. No matter what happens, no matter what changes occur, the numerical quantity called energy, which includes mass according to Equation (1), does not change. This idea is called the *Conservation of Mass – Energy*. Except for reactions involving atomic nuclei and elementary particles, changes in mass are too small to be observed. So in chemistry, this law is applied to mass and energy separately: in any chemical or physical process the total mass does not change and the total energy does not change. In other words, mass is conserved, and energy is conserved.

This conservation of energy leads to the dictionary definition of energy. The dictionary definition says that *energy* is a measure of the capacity to do work. Since energy is conserved, the concept of energy enables us to keep track of the capacity to do work.

Work is defined in science as a force applied over some distance. Since there are different kinds of forces, there appear to be different forms of energy (*e.g.* gravitational, electrical, chemical, nuclear, radiation, thermal, acoustic, and mechanical).

These different forms of energy can be classified into two types: kinetic energy and potential energy. *Kinetic energy* is the energy associated with the motion of an object. *Potential energy* is the energy associated with the position of an object.

The bottom line is that no one really understands what energy is, but it is a fundamental component of the universe, and we have ways of calculating energy. Whenever we calculate all the energy and add all the energy together, we always get the same number. This *conservation of energy* means that energy is neither created nor destroyed, but it can be converted from one form to another.

Energy is measured in units of joules, J. *Joule* rhymes with *rule* and was the name of an English physicist, who established that the energy of a system can be changed by heating and doing work. This idea, which now is known as the *First Law of Thermodynamics* is expressed as

$$\Delta E = q + w \tag{2}$$

where ΔE is the change in energy of the system, q is the energy transferred into the system, and w is the work that is done on the system.

Exploration: Measuring the Amount of Energy

Fig. 1. Transferring energy to water by heating.

As shown Fig. 1, an iron rod is heated over a Bunsen burner. It then is plunged into water. The temperature of the water rises from 18.1°C to 27.4°C, at which point the iron rod and water are at the same temperature. This exploration will determine the amount of energy transferred from the iron rod to the water.

1. What is the change in temperature of the water caused by heating the water with the hot iron bar?

2. If the beaker had half as much water in it, do you think that the temperature change would be the same, twice as large, half as large, or some other amount? Explain.

3. If the beaker had some other liquid in it, *e.g.* pancake syrup or honey, rather than water, do you think the temperature change would be the same or different? Explain.

Information

The specific heat capacity of a substance is the amount of energy required to increase the temperature of 1 g of that substance by 1°C. Specific heat capacities for some substances are given in Table I.

Table I. Specific Heat Capacities

Symbol or Formula	Substance	Specific Heat Capacity $J / g°C$
Fe	Iron	0.451
Al	Aluminum	0.902
$H_2O(l)$	Liquid water	4.184
$HOCH_2CH_2OH$	Ethylene glycol	2.42

4. Considering the units of specific heat capacity, what other information is needed to calculate the amount of energy transferred to a substance by heating?

5. Transform your answer to Question 4 into a mathematical equation using the following symbols: ΔT = change in temperature, c_s = specific heat capacity, m = mass, q = energy transferred.

Application

1. If there is 500. mL of water in the beaker in Fig. 1, how much energy was transferred from the iron rod to the water?

2. If this same amount of energy were transferred to 250. mL of water, what would the change in temperature of the water be?

3. If this same amount of energy were transferred to 500. mL of ethylene glycol, what would the change in temperature be?

4. Determine the final temperature of a 110 g aluminum cup when heated by 1.1 kJ of energy. The cup initially was at 23°C.

Got It!

1. Identify which will be hotter if you heat equal masses of iron and aluminum with 1.0 kJ of energy starting at room temperature. Explain.

2. You are holding an Al rod in one hand and a Fe rod in the other. The rods have the same mass and initial temperature (273 K). Assuming the rate of heat transfer is the same for both, which rod will reach your body temperature first? Explain.

06-2. Enthalpy: How much energy is produced by a chemical reaction under different conditions?

Energy is stored in molecules in different forms. Molecules have kinetic energy stored in the motion of the atomic nuclei and electrons, and they have potential energy stored in the electrostatic interactions involving all the nuclei and electrons.

When a chemical reaction occurs, reactant molecules are transformed into product molecules and the amount of energy stored must change because the molecules now are different. If the product molecules have more energy than the reactant molecules, then energy must be absorbed in the reaction. If the reactant molecules have more energy than the product molecules, then energy is released in the reaction.

Exploration

Fig. 1 shows a diagram of a calorimeter, which is a device for measuring the amount of energy that is transferred during a chemical reaction, that can be operated in either a constant volume mode (A) or in a constant pressure mode (B). In the constant volume mode, a bar is inserted to hold the piston in place. In the constant pressure mode, a pressure is applied that resists a change in volume.

Fig. 1. Combustion in a calorimeter operated at (A) constant volume and (B) constant pressure. The calorimeter heat capacity is 25.370 kJ/°C.

Two experiments are conducted with liquid isooctane, which is the ingredient in gasoline from which the isooctane rating is deived: one in the constant volume mode and one in the constant pressure mode. In each experiment 11.42 g of liquid isooctane (C_8H_{18}, 114.2 g/mol) is burned completely in excess oxygen to produce $CO_2(g)$ and $H_2O(g)$. The change in temperature of the calorimeter is measured in each case. The results of these experiments are given in Table I.

Table I. Results of Calorimetry Experiments with Isooctane.

Experiment	Sample mass	Initial temp	Final temp	ΔT
#1 constant V	11.42 g	21.560 °C	41.607 °C	
#2 constant P	11.42 g	23.693 °C	43.695 °C	

1. Write the balanced reaction equation for the combustion of liquid isooctane to produce gaseous carbon dioxide and water.

2. Calculate the temperature change of the calorimeter for experiments #1 and #2, enter the values you obtain in the last column of Table I, and include both values and units.

3 Use the heat capacity of the calorimeter, which is given in the figure caption, to calculate the energy transferred to the calorimeter in experiment #1 and experiment #2. Show your work below, and enter your results in Table II, including both values (5 sig figs) and units. Also enter the moles of isooctane consumed in each experiment.

Table II. Summary of the Calorimetry Experiments with Isooctane.

Experiment	Isooctane consumed (moles)	Energy transferred, q
#1 constant V		
#2 constant P		

4. In what ways were experiment #1 and experiment #2 different?

5. Since energy must be used to move the piston and consequently that energy is not available to increase the temperature of the calorimeter, which experiment and data in Table II give you the total change in energy of the reacting system (oxygen plus isooctane)? Explain.

6. The energy used to move the piston against the external pressure is called *work*. According the First Law of Thermodynamics, $\Delta E = q + w$, where ΔE is the change in energy of the system, q is the energy transferred, and w is the work done. In terms of the First Law of Thermodynamics:
(a) Did the energy of the reacting system increase or decrease in experiments #1 and #2? Explain.

(b) Is the sign of q positive or negative in these experiments? Explain. Go back and add the sign to the q values in Table II.

7. Use the data in Table II for the two experiments to determine a value for the work that was done by moving the piston in experiment 2. Show your work below.

Information

If energy, q, is transferred into a system or work, w, is done on a system, then the energy of the system increases. The change in energy of the system, ΔE, is given by Equation 1, which is called the *First Law of Thermodynamics*.

$$\Delta E = q + w \tag{1}$$

If energy is transferred out of the system or work is done by the system, then the energy of the system decreases. Equation 1 also describes this situation, except then q and w are negative ($q < 0$ and $w < 0$) because the energy of the system, E, decreases.

Work is done when a force moves some object through some distance. If nothing moves, then no work is done, $w = 0$, and

$$\Delta E = q_v. \tag{2}$$

The subscript in q_v indicates that energy is transferred with the volume constant and no work being done.

Physical processes and chemical reactions in plants, animals, laboratories, and our environment generally do not occur in closed containers with a constant volume. Instead they occur in contact with the atmosphere at constant pressure. If there is a change in volume, the atmosphere has to be pushed back. Work is done in pushing back the atmosphere, and this work is called expansion work or PV work.

Expansion work is determined by the change in volume and the pressure, P, resisting the expansion. The change in volume is given by $\Delta V = V_{final} - V_{initial}$. When $V_{final} > V_{initial}$, work has been done by the system, so the energy of the system must decrease, and w in equation (1) is negative.

$$w_p = - P \, \Delta V \tag{2}$$

The subscript in w_p indicates that only expansion work is done at constant pressure.

Since generally processes occur at constant pressure and not constant volume, the energy transferred between a system and its surroundings at constant pressure is a very important quantity and is given a special name. When the only work done is expansion work, the energy transferred at constant pressure is called the change in enthalpy of the system, ΔH.

$$q_p = \Delta E - w_p = \Delta H \tag{3}$$

The data in Table II for Experiment #1 gives you the change in energy, ΔE, for the combustion of isooctane, and the data for experiment #2 gives you the change in enthalpy, ΔH. They are different because of the expansion work that calculated in response to Question 7, from Equation (3) $\Delta E - \Delta H = w_p$.

Got It!

Gasohol, which is being used as an automotive fuel, is a mixture of ethanol and gasoline. The alcohol releases energy in a combustion reaction with oxygen.

$$CH_3CH_2OH(l) + 3O_2(g) \rightarrow 2CO_2(g) + 3H_2O(g)$$

The following data was obtained when 4.61 g of ethanol was burned in the calorimeter shown in Figure 1.

Table III. Results of Calorimetry Experiments with Ethanol.

Experiment	Sample mass	ΔT
constant V	4.607 g	5.412°C
constant P	4.607 g	5.392°C

Use these data to calculate the following quantities.

1. The molar enthalpy of combustion of ethanol.

2. The total change in energy of the reacting system when one mole of ethanol reacts with completely with three moles of oxygen.

3. The work done on the movable piston in the constant pressure experiment.

Application

Heating oil and liquefied petroleum gas (LPG) are two fuels that are commonly used for heating buildings. Use the calorimetry data in Table IV to compare these fuels in terms of the quantities below.

The data were obtained using a constant pressure calorimeter with a heat capacity of 50.745 kJ/°C. Assume heating oil is $C_{16}H_{34}$ (molar mass 226.00 g/mol, density 0.80 g/mL) and LPG is C_3H_8 (molar mass 44.10 g/mol, density 0.51 g/mL).

Table IV. Results of Calorimetry Experiments with Heating Oil and Liquefied Petroleum Gas

Substance	Sample Mass	ΔT
Heating Oil	2.261 g	2.128°C
LPG	4.411 g	4.020°C

1. The molar enthalpy of combustion in units of kJ/mol.

2. The fuel value in units of kJ/g.

3. The energy density in units of kJ/mL.

4. The environmental value in kJ/mol CO_2 produced.

5. The economic value in MJ/$ if both cost $2.50/gal.
   (1 gal = 3.785 L)

07-1. Photoelectron Spectrum of Argon: How do we learn about the electronic structure of atoms?

Before beginning this activity, you need to read your textbook to learn about electromagnetic radiation, the photoelectric effect, the Bohr model of the hydrogen atom, and the quantum numbers n, l, and m_l that are used to identify the electron atomic orbitals and energy levels in atoms. The relevant information that you should read about in detail is summarized below.

The lowest energy state of an atom or molecule is called the *ground state*. The absorption of electromagnetic radiation by matter can excite atoms and molecules to higher energy states and even ionize electrons.

Electromagnetic radiation has both wave properties and particle properties. The wave properties include a frequency, ν, wavelength, λ, and the speed of light, c. These quantities are related by the following equation.

$$\nu\lambda = c \qquad (1)$$

The particle properties include momentum and the fact that energy is delivered in packets called photons. The energy of a photon is given by

$$E_{photon} = h\nu, \qquad (2)$$

and the momentum of a photon is given by

$$p = h/\lambda \qquad (3)$$

Visible light is one region of the electromagnetic spectrum. Visible light extends from a wavelength of 400 nm to 700 nm. X-rays form another region of the electromagnetic spectrum. X-rays have very short wavelengths and high photon energies.

In photoelectron spectroscopy, photons are used to ionize electrons from atoms, molecules, and solids. This phenomenon was first discovered in the early 1900s and is called the photoelectric effect.

A photoelectron spectrum is a plot or graph of the kinetic energy of the emitted electrons on the x-axis, and the number of electrons emitted per second at each kinetic energy on the y-axis.

Photoelectron spectra of argon are shown in Figures 1 and 2. Such spectra can be obtained by ionizing argon atoms with x-ray photons that have an energy of 3500 eV.

Electron volts (eV) is the energy unit that is preferred by scientists who use this technique because the joule unit is too large (1eV = 1.602×10^{-19} J).

The key equation used to analyze photoelectron spectra is obtained by considering the conservation of energy. The energy of the photon is used to overcome the binding energy of the electron in the atom, molecule, or solid, and any excess energy goes into the kinetic energy of the emitted electron.

$$E_{photon} = E_{binding} + E_{kinetic} \tag{4}$$

Electron atomic orbitals are identified by quantum numbers. The principal quantum number, n, has integer values (n = 1, 2, 3, 4...). As n increases, the energy of the electron increases, and the electron on average is further away from the nucleus and less tightly bound to it. For each value of n, the azimuthal quantum number, l, has integer values that range from 0 up to a maximum of n − 1.

Are You Prepared to Explore?

1. What is the wavelength of an x-ray photon that has energy of 3500 eV?

2. What happens in the photoelectric effect?

3. What quantities are plotted on the x and y axes of a photoelectron spectrum?

4. What happens to the energy of an electron in an atom as the value of the principal quantum number increases?

Exploration

Fig. 1 Photoelectron spectrum of argon from 0 to 500 eV.

Fig. 2 Photoelectron spectrum of argon from 3000 to 3550 eV.

1. Peaks in the above photoelectron spectra occur at electron kinetic energies of 295, 3175, 3250, 3470, and 3485 eV. Label each peak in Figures 1 and 2 with the appropriate value of the electron kinetic energy: 295, 3175, 3250, 3470, and 3485 eV.

2. Rearranging Equation (4) produces Equation (5) that shows electron binding energy can be obtained by subtracting the electron kinetic energy from the photon energy. Using the data in Figures 1 and 2, determine the electron binding energies in argon and write the values in Table I below.

$$E_{binding} = E_{photon} - E_{kinetic} \tag{5}$$

65

3. Check that the sum $E_{kinetic} + E_{binding}$ in each column of Table 1 adds up to 3500 eV. Explain why this sum must be 3500 eV.

4. The different peaks in the photoelectron spectrum are produced by ionizing electrons with different binding energies. The most tightly bound electrons, those with the largest binding energy, are in orbitals that have the smallest values of n and l. Write the values of n and l in Table I to assign the peaks in the photoelectron spectrum. The possible values are (n,l) = (1,0), (2,0), (2,1), (3,0), and (3,1).

Table I. Argon Photoelectron Data

$E_{kinetic}$(eV)	3485	3270	3250	3175	295
$E_{binding}$(eV)					
n value					
l value					

5. Using the data in Table I, complete the energy level diagram below to show the binding energies of electrons in argon. Label each energy level with the value of the binding energy and the values for the quantum numbers n and l. The lowest energy level is done for you.

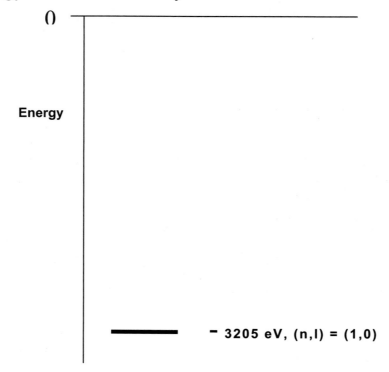

Information

The photoelectron spectrum and energy level structure can be understood in terms of a shell model for the atom. Electrons in the n=1 shell are closest to the nucleus and are strongly bound by the +18 charge of the argon nucleus. Electrons in the n=3 shell are furthest from the nucleus and are only weakly bound. Electrons in the n=2 shell are intermediate between these two. There are 3 shells with n = 1, 2, and 3, but 5 peaks are observed in the photoelectron spectrum. The n=2 and n=3 shells are split into subshells with each subshell identified by a different value for l.

Application - 1

Complete the diagram below to illustrate the shell model for argon as described in the preceding *Information* section. Each subshell outside the +18 nuclus is indicated by a solid line. Label subshells with the quantum numbers n and l: (n,l) = (1,0), (2,0), (2,1), (3,0), and (3,1). Add the 18 electrons as small dark circles. Put 2 electrons in each of the subshells that have l = 0, and put 6 in each of the subshells that have l = 1.

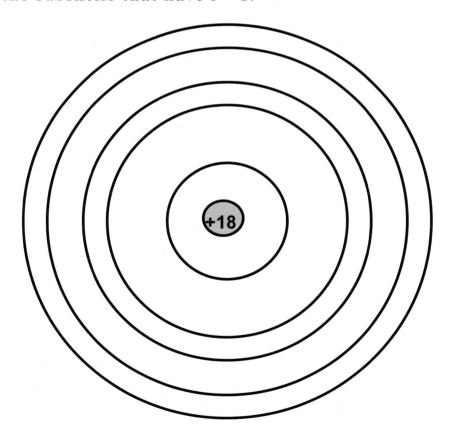

Research

Use your text as a resource and explain why there are 2 and only 2 electrons in each of the shells with $l = 0$, and 6 electrons in each of the shells with $l = 1$.

What do you predict?

Potassium has 19 protons and 19 electrons. The additional electron, compared to argon, goes in the $n = 4$, $l = 0$ subshell.

1. Will the radius of a potassium atom be larger or smaller than that of argon? Explain in terms of the shell model diagram that you drew in *Application – 1* above.

2. Will it require more or less energy to remove an electron from potassium compared to argon? Explain in terms of the shell model diagram that you drew in *Application – 1* above. Remember, an electron is bound in an atom because it is attracted to the positively charged nucleus even though it is repelled by the other electrons, and the net result is a balance between the attraction by the nucleus and the repulsion by the other electrons.

Application - 2

1. Based on your experience with analyzing the argon photoelectron spectrum, construct a photoelectron spectrum for krypton. Plot electron binding energy on the x-axis rather than the kinetic energy of the ionized electron since the photon energy used for ionization is not specified. The atomic number of krypton is 36, and the electron binding energies are 14326, 1921, 1704, 293, 218, 94.5, 27.5, 14.1 eV.

2. Construct an energy level diagram for krypton similar to the one you completed for argon.

3. Draw a shell model for krypton similar to the one shown in Application – 1 for argon.

07-2. Periodic Trends in Properties of Elements: How much can you learn from the Periodic Table?

With knowledge of the shell structure of atoms and electron configurations, you can understand why different atoms are in the positions they are in the Periodic Table. You also can explain trends in some properties of the elements. In this activity, you examine trends in atomic radius, ionization energy, and electron affinity.

Exploration – 1: The Size of Atoms

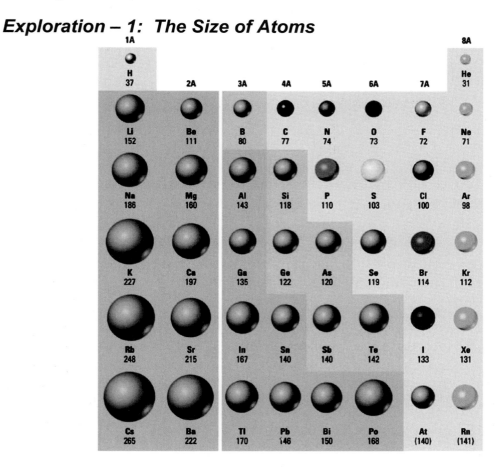

Fig. 1. Atomic radii of the main group elements in pm (10^{-12} m).

1.1. According to the information in Figure 1, does the atomic radius increase or decrease as the atomic number increases across a period (row) in the Periodic Table?

71

1.2. In going from one element to the next across a period, does the number of protons and the number of electrons increase or decrease?

1.3. Since opposite electrical charges attract each other, an increase in the number of protons in the nucleus should increase the electron – nucleus attraction, pull the electrons closer to the nucleus, and make the atom smaller. Since like charges repel each other, an increase in the number of electrons should increase the electron – electron repulsion, push the electrons apart, and make the atom larger. In view of you answer to Question 1.1, which effect evidently dominates in going from one atom to the next across a period: electron – nucleus attraction or electron – electron repulsion? Explain.

1.4. According to the information in Figure 1, what happens to the atomic radius down a group (column) in the Periodic Table?

1.5. In going from one element to the next down a group, does the number of protons and the number of electrons increase or decrease?

1.6. In view of you answer to Question 1.4, which effect evidently dominates in going from one atom to the next down a group: electron – nucleus attraction or electron – electron repulsion? Explain.

1.7. Use the shell structure of atoms to explain why electron – nucleus attraction dominates and makes atoms smaller in going from one atom to the next across a period. Include in your explanation the fact that electrons are added to the same shell in going across a period.

1.8. Use the shell structure of atoms to explain why electron-electron repulsion dominates and makes atoms larger in going from one atom to the next down a group. Include in your explanation the fact that electrons are added to different shells and the distance of the outer electrons from the nucleus increases in going down a group.

Exploration – 2: Ionization Energies

Fig. 2. First ionization energies for elements in the first six periods.

2.1. According to the information in Figure 2, does the ionization energy generally increase or decrease across a period in the Periodic Table?

2.2. Explain whether or not the changes in ionization energy and atomic radii across a period are consistent with the following statement: an increase in the electron-nucleus attraction increases the energy required to remove an electron from an atom and pulls electrons closer to the nucleus.

2.3. According to the information in Figure 2, does the ionization energy increase or decrease down a group in the periodic table?

2.4. Explain whether or not the changes in atomic radii and ionization energies down a group are consistent with the following statement: an increase in the distance of an electron from the nucleus and an increase in electron-electron repulsion decreases the energy required to remove an electron from an atom and leads to an increase in the atomic radius.

Application

Using only a Periodic Table and your knowledge of periodic trends in atomic size and ionization energies that you gained by answering the above questions, make the following predictions based on their positions in the Periodic Table without looking at Figures 1 and 2.

1. For each pair, predict which one is larger:
   N or O, H or He, Br or Cl, Na or Mg, Si or C, Ge or P

2. For each pair, predict which one has the higher ionization energy: H or He, Ne or Ar, Na or Ne, F or Cl, K or Cs

Exploration – 3: Electron Affinities

The *ionization energy* of an atom is defined as the energy required to ionize an atom, corresponding to the following reaction equation.

$$X \rightarrow X^+ + e^- \qquad\qquad (1)$$

By analogy, research scientists define the *electron affinity* of an atom as the energy required to ionize the atomic anion, corresponding to the following reaction equation. By this definition, the electron affinity is a positive quantity.

$$X^- \rightarrow X + e^- \qquad\qquad (2)$$

In some textbooks, the electron affinity is defined as the energy associated with the following electron attachment reaction.

$$X + e^- \rightarrow X^- \qquad\qquad (3)$$

By this definition, the electron affinity is a negative quantity because electron attachment, Equation 3, is just the reverse of anion ionization, Equation 2. Energy is required to ionize the anion ($\Delta E>0$ for Equation 2), this energy is returned when the anion is formed ($\Delta E<0$ for Equation 3). The electron affinities defined by Equation (2) and used by research scientists are listed in Figure 3. If your textbook lists negative electron affinities, they are defined by Equation (3) not Equation (2).

Electron Affinities (kJ/mol)

1A (1)	2A (2)	3A (13)	4A (14)	5A (15)	6A (16)	7A (17)	8A (18)
H 73							He 0
Li 60	Be 0	B 27	C 122	N 0	O 141	F 328	Ne 0
Na 53	Mg 0	Al 43	Si 134	P 72	S 200	Cl 349	Ar 0
K 48	Ca 2	Ga 30	Ge 119	As 78	Se 195	Br 325	Kr 0
Rb 47	Sr 5	In 30	Sn 107	Sb 103	Te 190	I 295	Xe 0

Fig. 3. Electron affinities of main group elements.

3.1. Which group has elements with the largest electron affinities?

3.2. Which two groups have elements with electron affinities that are zero or near zero?

3.3. Draw a shell model diagram for Ne and Ne⁻, and use the diagram to explain why the electron affinities of the noble gases all are zero. In your explanation include the effects of the distance of the outer electron from the nucleus, electron-electron repulsion, and electron-nucleus attraction.

3.4. Draw a shell model diagram for F and F⁻, and use the diagram to explain why the electron affinities of the halogens are all very large. In your explanation include the effects of the distance of the outer electron from the nucleus, electron-electron repulsion, and electron-nucleus attraction.

3.5. The electron affinity of nitrogen is zero. Compare the electron configurations of N and N⁻ and explain in terms of electron-electron repulsion why N⁻ is not stable.

3.6. The electron affinity of beryllium is zero. Compare electron configurations and shell model diagrams of Be and Be⁻ and explain in terms of electron-electron repulsion and distance of the electrons from the nucleus why Be⁻ is not stable.

08-1. Lewis Structures: What strategy produces the correct Lewis structure every time?

Atoms combine to form molecules because molecules are more stable than atoms. *More stable* means that molecules have a lower energy than the separated atoms.

The attractive force that holds a pair of atoms together in a molecule is called a *bond*. One picture of bonding views a bond as one or more pairs of electrons between two nuclei. This simple model works very well. The negative charge of the two electrons attracts each nucleus, thereby effectively attracting the two positively charged nuclei to each other. This attraction holds the two atoms together.

Bonds are formed by the outer least tightly bound electrons of atoms. These electrons are called the *valence electrons*. The valence electrons generally come from the highest energy s and p atomic orbitals.

G.N. Lewis figured out a way to determine how the valence electrons in a molecule arrange themselves to form bonds between atoms. Lewis reasoned that since the noble gases are very stable and have 8 electrons (*an octet*) in their outermost or valence shell, then stable molecules would be produced if one or more pairs of electrons were shared between the bonded atoms to give each atom an octet of electrons. Bond formed by atoms sharing pairs of electrons are called *covalent bonds*.

In this activity, you will learn how to draw Lewis structures, determine how many pairs of nonbonding electrons there are in a molecule, and identify whether the bonds are single, double, or triple covalent bonds.

Figure 1 provides examples of Lewis structures for three molecules showing single bonds, a carbon-carbon triple bond, a carbon-oxygen double bond, and lots of nonbonding electron pairs.

Fig. 1. Lewis Structures

Exploration – A Strategy for Drawing Lewis Structures

Step	Examples
1. Count the total number of valence electrons in the molecule or ion. The Group number (IA – VIIIA) in the column heading of the Periodic Table gives you the number of valence electrons for a neutral atom. Subtract the ion charge for a cation, add the ion charge for an anion.	PCl_3 has 26 valence electrons: 5 from P and 3 x 7 = 21 from Cl. NO_2^- has 18 valence electrons: 5 from N, 2 x 6 = 12 from O, and 1 more because it is a −1 anion.
2. Decide which atoms are bonded to each other, and use a pair of electrons, represented by a line to connect the atomic symbols for these atoms. The central atom in a structure usually is the one written first in the molecular formula, is the most electropositive, and can form the most bonds with other atoms.	Cl—P—Cl 　　Cl O—N—O
3. Place the remaining electrons as lone pairs (aka nonbonding pairs) around each atom to satisfy the octet rule starting with the atoms on the outside. H only gets 2 electrons since H only has room for 2 electrons in its valence shell.	
4. Place any leftover electrons on the central atom even if it will give that atom more than an octet. Period 3 elements and higher can accommodate more than 8 electrons.	
5. If the number of electrons around the central atom is less than 8, use one or more lone pairs from an outside atom to form double or triple bonds until all the atoms have an octet of electrons. A double bond has 2 shared pairs, and a triple bond has 3 shared pairs.	
6. Draw any equivalent resonance structures. The two bonds between N and O in the nitrite anion are equivalent. To represent this fact, draw two equivalent Lewis structures, and nitrite is said to be a resonance hybrid of these two structures.	
7. Calculate and write the formal charge on each atom. For each atom, the formal charge = the number of valance electrons contributed by the atom in Step 1 – the number of lone pair electrons on that atom – half the number of bonding electrons associated with that atom.	

1. How can you determine the number of valence electrons that an atom has from its position in the Periodic Table?

2. Given a molecular formula, which atom usually is the central atom in the Lewis structure?

3. What do you do in drawing a Lewis structure if each atom has been assigned an octet of electrons, but you have not used all the valence electrons?

4. What do you do in drawing a Lewis structure if you have used all the valence electrons, but each atom does not have an octet of electrons?

5. Why might it be necessary to draw equivalent resonance structures to represent a molecule?

6. How is the formal charge of an atom in a Lewis structure determined?

7. Sometimes more than one Lewis structure can be constructed following this procedure. Use your textbook as a resource and describe below how the formal charge can be used to identify which structure has the lowest energy and therefore is probably the best representation of the most stable state (the ground state) of the molecule.

Application

1. Draw Lewis structures for the following molecules. Include formal charges in your diagrams.

    hydrofluoric acid, HF

    ammonia, NH_3

    phosphate, PO_4^{3-}

    butane, C_4H_{10}

    ethanol, CH_3CH_2OH

    carbon dioxide, CO_2

    acetylene, C_2H_2

    Chloroethene, C_2H_3Cl

    Chlorine trifluoride, ClF_3

    nitrate, NO_3^-

2. Draw Lewis structures for carbonyl sulfide with three possible connectivities: O-C-S, C-O-S, and C-S-O. Use the formal charges on the atoms as a criterion to predict which molecule is the most stable.

Got It!

Use Lewis structures and formal charge to predict whether $S-C\equiv N^-$ or $S-N\equiv C^-$ is the more stable anion.

08-2. Electronegativity and Bond Properties: Which atom attracts more electron density?

In covalent bonds, pairs of electrons are shared between two atoms, but the electrons are not shared equally unless the two atoms are the same. In molecules such as oxygen and nitrogen, O_2 and N_2, where both atoms are the same and the sharing is equal, the bond is said to be *nonpolar*. When the atoms are different and the electrons are pulled more toward one atom than the other, the bond is said to be *polar*. It even is possible for one atom to pull the valence electrons away from the other completely to form ions. In that case the bond is said to be *ionic*.

Electronegativity is the ability of an atom to attract electrons away from another atom in a molecule. Linus Pauling first proposed the concept of electronegativity and assigned an electronegativity value to each atom, giving fluorine, which is the most electronegative atom, a value of 4.0. Electronegativity can be thought of as an average of the *ionization energy* and *electron affinity* of an atom. The ionization energy is a measure of the strength of an atom in keeping an electron, and the electron affinity is a measure of the strength of an atom in attracting another electron. If both of these are large, then the electronegativity is high. If both of these are small, then the electronegativity is low.

Exploration – 1. Electronegativity of the Elements

Fig. 1. Electronegativity of the elements.

81

1.1. Compare the ionization energies and electron affinities in the activity on periodic trends with Figure 1. In which part of the Periodic Table are elements with high ionization energies, high electron affinities, and high electronegativities found?

1.2. In which part of the Periodic Table are elements with low ionization energies, low electron affinities, and low electronegativities found?

1.3. Which are more electronegative, metals or nonmetals?

1.4. The polarity of a bond is quantified by a value called the dipole moment, μ. The larger the dipole moment, the more polar the bond. What relationship between bond polarity and the difference in electronegativity of the elements (ΔEN) can you identify from the data in Table 1?

Table 1. Measures of Bond Polarity

Bond	μ in debye	EN values	ΔEN
F – F	0	F = 4.0	0
H – Br	0.8	H = 2.1 Br = 2.8	0.7
H – F	1.9	H = 2.1 F = 4.0	1.9
Na – F	8.2	Na = 0.9 F = 4.0	3.1

Information

A larger difference in electronegativity between two atoms produces a more polar bond. The shift from nonpolar covalent bonds to ionic bonds can be regarded as a continuum as shown in Figure 2. Polar covalent bonds are in the middle, and slightly and very polar covalent bonds are on either side of the middle. Ions and an ionic bond generally are formed when the electronegativity difference between the atoms is 2.0 or larger.

Fig. 2. Bond character, dipole moments, and electronegativity differences for four substances.

Got It!

For each of the following pairs of bonds, identify from the electronegativity data in Figure 1, the atom that has a small positive charge ($\delta+$), the one that has a small negative charge ($\delta-$), and indicate which bond is more polar.

1.1. Cl – F and Cl – Br

1.2 Si – H and C – H

1.3. P – Cl and S – Cl

1.4. C – O and C – N

1.5. C – O and Si – O

Exploration – 2. Bond Lengths and Strengths

Table 2. Bond Lengths and Strengths

Bond	Bond Length (pm)	Bond Strength (kJ/mol)
C – C	154	356
C = C	134	598
C $\equiv$ C	121	813
N – N	140	160
N = N	120	418
N $\equiv$ N	110	946

2.1. As the multiplicity of the bond increases from single to double to triple, what happens to the bond length?

2.2. As the multiplicity of the bond increases from single to double to triple, what happens to the bond strength?

2.3. Does the electron density between the two nuclei increase or decrease as the bond multiplicity increases, and why does this change affect the bond length and bond strength?

2.4. Why are N – N bond lengths shorter than C – C bond lengths? Explain in terms of the relative sizes of N and C atoms: see the activity on periodic trends for atomic radii.

Got It!

For each of the following pairs of bonds, predict which one is shorter and has the higher bond energy. Explain your rationale in each case by using the ideas of atomic size and the electron density between the atoms.

2.1. C – C or C – H

2.2. C = O or C = N

2.3. C – Cl or C – Br

2.4. C $\equiv$ N or C = N

09-1. VSEPR Model: How can the geometry or shape of a molecule be predicted?

The geometry or shape of a molecule is important in determining its reactivity, its function in biological systems, and the physical and chemical properties of the substance. The valence-shell electron-pair repulsion model (VSEPR) provides a simple and reliable method for predicting the shapes of molecules and polyatomic ions. The VSEPR model is based on the idea that regions of high electron density in a molecule repel each other. These regions include the bonds and the nonbonding or lone pairs of electrons.

Exploration – A Strategy for Determining Molecular Shapes

Step	Examples

1. Draw the Lewis structure.
See the activity on Lewis structures.
	H—N̈—H
	      |
	     H

2. Count the number of bonds and lone pairs around each atom. Count double and triple bonds as 1.	In ammonia: 3 bonds + 1 lone pair = 4
3. Arrange the bonds and lone pairs to maximize their separation, which minimizes the electron – electron repulsion energy.	Four regions of electron density are furthest apart if they have a tetrahedral shape. **Tetrahedral electron geometry**
4. Add the atoms to positions that are consistent with the Lewis structure, and keep the lone pairs (if there is more than one lone pair) as far apart as possible.	
5. Determine the molecular shape from the positions of the atoms, ignoring the lone pairs.	Ammonia is a trigonal pyramid. **Trigonal pyramidal molecular geometry**

1. In Step 1, you need to determine the Lewis structure. Write below the procedure that you follow to draw a correct Lewis structure.

2. In determining the electron geometry, why are the bonding and lone pair electrons spaced as far apart as possible?

3. According to Step 4, if you have two lone pairs and four atoms around a central atom, would you position the lone pairs at 90° or 180° to each other? Explain.

4. There are 5 possible arrangements or geometries for the bonds and lone pairs of electrons. Use your textbook as a reference to help you complete Table 1, below.

Table 1. Electron Pair Geometries

Number of bonds and lone pairs.	Electron Geometry	Sketch of the Electron Geometry
2	Linear	
3	Trigonal planar	
4	Tetrahedral	
5	Trigonal bipyramidal	
6	Octahedral	

5. In Step 4 you add the atoms to the electron geometry, and in Step 5, you determine the molecular geometry or shape by looking at the positions of the atoms and ignoring the lone pairs. Use your textbook as a resource to help you sketch the molecular shape for the following examples where the molecular geometry is given.

(a) 3 atoms, linear, CO_2

(b) 3 atoms, bent or angular, SO_2

(c) 4 atoms, trigonal planar, $CO_3{}^{2-}$

(d) 4 atoms, trigonal pyramidal, NH_3

(e) 4 atoms, T-shaped, ClF_3

(f) 5 atoms, tetrahedral, CH_4

(g) 5 atoms, seesaw, SF_4

(h) 5 atoms, square planar, XeF_4

(i) 6 atoms, square pyramidal. BrF_5

(j) 6 atoms, trigonal bipyramidal, PF_5

(k) 7 atoms, octahedral, SF_6

Got It!

For the following molecules: draw the Lewis structure, and use the VSEPR model to identify the electron geometry and the molecular geometry.

Molecule	Lewis Structure	Electron Geometry	Molecular Geometry
H_2O			
I_3^-			
$SeCl_6$			
SbF_5			
H_2CO			
SeO_3^{2-}			
SiF_4			
KrF_4			
SF_4			
ICl_3			
BrF_5			

09-2. Hybrid Atomic Orbitals: How can atomic orbitals describe all the different electron geometries found in molecules?

Review of Prerequisite Knowledge

The forces that hold molecules together are electrostatic. It is as simple as opposite charges (+ and −) attract and like charges (+ and + or − and −) repel. However, when the theory of electrostatics was applied, it could not account for any of the properties of atoms and molecules or even for their existence. Classical electrostatic theory says that a proton and an electron should attract each other, come together, and the charges cancel each other, not exist as a hydrogen atom.

As a result of this situation, a new theory was invented. This theory is known as *Quantum Mechanics* or *Schrödinger's Wave Mechanics*. Schrödinger figured out how to find the waves that describe electrons in atoms and molecules. Application of Schrödinger's ideas to the hydrogen atom produced the electronic wave functions Ψ_{nlm} for the hydrogen atom. Each wave function is identified by unique values for the quantum numbers n, l, and m_l. Any question about the hydrogen atom can be answered by using its wave functions. These functions are called atomic orbitals. Pictures of these functions show the regions of space where the function has large positive or negative values.

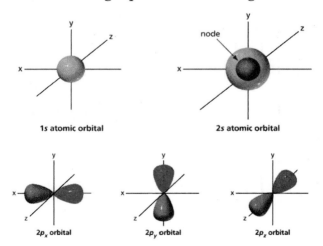

Fig. 1 Pictures of the 1s, 2s, and 2p atomic orbitals.

The square of the wave function Ψ_{nlm}^2 is the electron probability density. It gives the probability density of finding an electron at different points in space.

In a chemical bond there is a high density of electrons between the two atomic nuclei. It is the negative charge from the electrons that pulls the two positively charged nuclei together to form the bond. However, the atomic p orbitals don't describe many of these bonds very well because they point in the wrong directions. Consider the case of methane shown below.

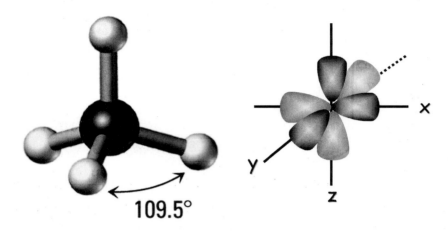

109.5°

Fig. 2 Methane (CH₄) and the three atomic p orbitals.

Figure 2 shows that methane has a tetrahedral shape, and that the three atomic p orbitals of carbon are directed along the x, y, and z axes in both directions. The bonds in methane lie at an angle of 109.5° to each other, while the x, y, and z axes and atomic p orbitals are at angles of 90° to each other. The p orbitals are off to one side of the bonds or the other and extend in both directions along the x, y, and z axes. The p orbitals therefore do not describe where the electrons are in the methane molecule.

New orbitals are needed. These new orbitals are formed by combining the s and p orbitals The orbitals made from this combination are called *hybrid orbitals*. The hybrid orbitals are designed to lie along the directions of the bonds because that is where the electrons are.

Five different types of hybrids can be made using combinations of s, p, and d orbitals. These five types of hybrid orbitals are shown in Figure 3.

Exploration: Hybridization of Atomic Orbitals

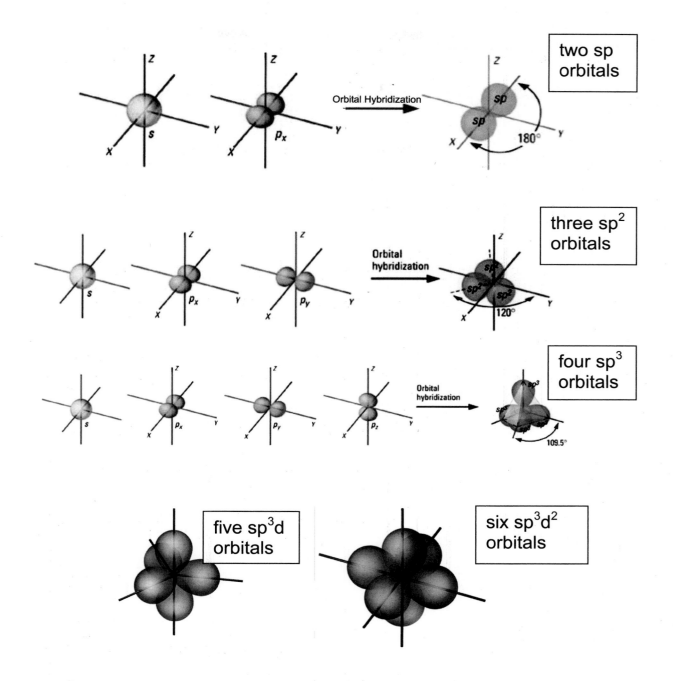

Figure 3 Formation of hybrid orbitals from s, p, and d atomic orbitals.

1. What are the names of the five types of hybrid orbitals shown in Figure 3? For example, one type is sp^3.

2. How many different hybrid orbitals are there of each type? For example, there are four sp^3 orbitals.

3. How many atomic orbitals and which ones must be combined to make each type of hybrid orbital? For example, four atomic orbitals (s, p_x, p_y, and p_z) combine to produce the four sp^3 hybrid orbitals.

4. In general, what is the relationship between the number of atomic orbitals used and the number of hybrid orbitals produced?

5. What is the angle between the hybrid orbitals for each type? For example in sp^3, the angle between the hybrid orbitals is $109°$.

6. Which set of hybrid orbitals in Figure 3 match the bond angles shown in Figure 2 for methane?

7. Figure 4 shows the bonds in acetylene C_2H_2. Which set of hybrid orbitals in Figure 3 matches the bond angles of 180° I acetylene?

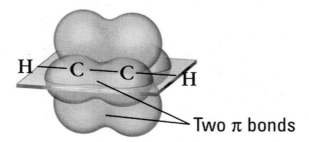

Fig.4 Bonding in acetylene. The lines represent bonds formed using carbon sp hybrid orbitals. The "clouds" represent bonds formed from the carbon atomic p_y and p_z orbitals on the two carbon atoms. The pair of p_y orbitals form one pi bond, and the pair of p_z orbitals form a second pi bond.

8. Figure 5 shows the hybrid orbitals used to describe the bonds in ethylene $CH_2=CH_2$. Which set of hybrid orbitals in Figure 3 matches the bond angles of 120° in ethylene?

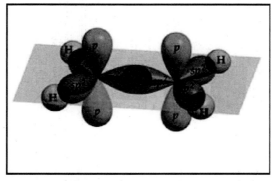

Fig. 5 Atomic and hybrid orbitals for ethylene. Note the p_z orbitals on the two carbon atoms overlap and form a pi bond just as they do in acetylene.

9. Why is it necessary to include the concept of hybrid orbitals in the description of molecules?

Information

Molecular orbitals and bonds are identified as being either sigma or pi using the Greek symbols σ and π, repectively. Sigma orbitals and bonds are formed from the overlap of atomic or hybrid orbitals along the axis or line connecting the two atomic nuclei. Pi orbitals and bonds are formed from the sideways overlap of parallel atomic p orbitals.

Application

1. List the number of sigma and pi bonds for each one of the following: methane, ethylene, and acetylene.

Compound	Sigma (σ) Bonds	Pi)(π) Bonds
Methane, CH_4		
Ethylene, C_2H_4		
Acetylene, C_2H_2		

2. Using your knowledge of the Lewis structure and the VSEPR model, complete the table below.

Molecule	Electron Geometry	Molecular Geometry	Central Atom Hybridization	Bond Angles
carbon tetrafluoride CF_4				
ozone O_3				
formaldehyde H_2CO				
carbonate anion CO_3^{2-}				
sulfur dioxide SO_2				
nitrate anion NO_3^-				

Got It!

Draw the Lewis structure of $CH_3CHCHCCH$, and identify the hybridization and bond angles for each carbon atom. In this condensed structural formula, the H atoms are bonded to the carbon on their left.

09-3. Organic Functional Groups: What determines the properties of complicated organic molecules?

A *functional group* is a combination of atoms bonded together in a characteristic pattern. The functional groups in an organic molecule are responsible for the chemical reactions that involve the molecule, its biological and medicinal activity, and the properties of the compound (*e.g.* solubility, vapor pressure, melting and boiling points). For example, *verapamil*, which is shown below, is used to treat irregular heartbeats and high blood pressure. It has the following functional groups: 2 aromatic rings, 4 ethers (R–O–R), a nitrile (–CN), an amine (–NR₂), and alkane linkages (–CH₂–).

For each functional group that you encounter in this activity, you need to remember its name and its characteristic bonding pattern. In a line drawing, like that shown for verapamil, each line represents a bond, and a carbon atom is implicit wherever a line ends or lines come together. Carbon atoms always have 4 bonds, so bonds to hydrogen atoms are implicit wherever 4 bonds to a carbon atom are not shown.

Verapamil
$C_{27}H_{38}O_4N_2$

What do you know about functional groups already?

1. After reading the introduction above, identify five functional groups that you know by name and label them in the diagram of verapamil.

Exploration: Some Common Functional Groups

Complete Tables I and II by entering the distinguishing characteristics where they are missing.

Table I. Some Common Functional Groups

Structure	Name	Example	Distinguishing Characteristics
C —— C	alkane	Propane, $CH_3CH_2CH_3$, a fuel	Carbon – carbon single bonds.
C ═══ C	alkene	Ethylene, CH_2CH_2, used to make polyethylene	
C ≡≡≡ C	alkyne	Acetylene, CHCH, used in welding torches	
(hexagon ring)	aromatic ring	Benzene, C_6H_6, a carcinogen	A hexagon of six carbons with alternating double and single bonds made equivalent by two resonance structures.
C — O — H	alcohol	Ethanol, CH_3CH_2OH, ingredient in beverages and gasoline	
C — X	halide	Carbon tetrachloride, CCl_4, X = any halogen used in dry cleaning.	
— N —	amine	Ephedrine, a decongestant	
C — O — C	ether	Diethylether, $(CH_3CH_2)O$, an anesthetic	
N ≡≡≡ C	nitrile	Acetonitrile, CH_3CN, an important solvent	
C ═══ O	carbonyl	Part of other groups shown in Table II.	

Table II. Carbonyl Containing Functional Groups

Structure	Name	Example	Distinguishing Characteristics
	aldehyde	Formaldehyde, CH_2O, preservative for biological specimens	a carbonyl group with a carbon and hydrogen attached
	ketone	Acetone, $(CH_3)_2O$, a product of fatty acid metabolism	
	carboxyl	Acetic acid CH_3CH_2COOH, an ingredient in vinegar	a carbonyl group with an OH and carbon attached
	acid halide. X = any halogen	Acetyl chloride, CH_3COCl, A very reactive compound used in synthesis	
	ester	Ethyl acetate, nail polish remover	
	amide	Nylon 6, a polymer used in fabrics, cords, and ropes	A carbonyl group with an amine group attached

Got It!

1. Para-aminobenzoic acid, PABA, is used in sun screen lotions. What are the three functional groups in this molecule?

2. What would you call this ingredient that is
   found in rhubarb: a carboxylic acid, a ketone, an
   alcohol, an ester, or an ether? Explain.

3. What are three functional groups present in aspirin?

4. What is different about these two steroids?

androstenedione   **testosterone**

5. Which of the following would <u>not</u> be a good hypothesis? Both of
   these molecules are narcotic pain relievers because they both have
   A) amine groups.
   B) similar structures.
   C) oxygen atoms in key positions.
   D) ester groups.

oxycodone

morphine

Making Connections

In addition to recognizing functional groups, you need to be able to recognize the local electron geometry and molecular geometry associated with each functional group by using the ideas developed in the activities on the VSEPR model and atomic orbital hybridization.

1. Complete the table below by identifying the local electron geometry and local molecular geometry around the atom specified.

Atom & Functional Group	Atomic Orbital Hybridization	Local Molecular Geometry
C atom in an alkane	sp^3	tetrahedral
C atom in an alkene		
C atom in an alkyne		
O atom in an alcohol		
O atom in an ether		
N atom in an amine		
C atom in a nitrile		
C atom in a carbonyl		

2. Answer the following questions, and for each one provide an explanation in terms of the local electron geometry and molecular geometry around the atoms relevant to the question.

(a) Which atoms in oxalic acid all lie in the same plane?

(b) Which atoms in p-aminobenzoic acid lie in the plane of the aromatic ring?

99

(c) Identify the atomic orbital hybridization of all the atoms in p-aminobenzoic acid.

3. Answer the following questions regarding verapamil

(a) What is the atomic orbital hybridization and local molecular geometry around the nitrogen atom of the amine group?

Verapamil
$C_{27}H_{38}O_4N_2$

(b) Do the two aromatic rings in lie in the same plane? Explain.

(c) What is the atomic orbital hybridization and local molecular geometry around the oxygen atoms?

(d) Do the all the carbon atoms in the chain connecting the two aromatic rings lie in the same plane? Explain.

10-1. Gases and the Ideal Gas Law: How much air does it take to fill a balloon?

What do you predict?

Identify whether you *agree* or *disagree* with the following predictions pertaining to the gas inside a can of hair spray. Assume all the liquid has been used, and only some residual gas consisting of hair spray and propellant molecules remain.

1. If the temperature changes from 15°C to 30°C, the pressure inside the can will double.

2. If half the mass of the gas is removed from the can, the pressure inside the can will decrease by a factor of 2.

3. The can contains fewer molecules than an identical can of hydrogen gas at the same temperature and pressure because the hair spray molecules and propellant molecules are larger than hydrogen molecules.

Information

The relationships among the pressure (P), volume (V), Kelvin temperature (T), and amount (moles, n) of a gas are expressed in a single equation called the *Ideal Gas Law*. This equation includes the gas constant (R).

$$PV = nRT \tag{1}$$

This equation can be used to calculate P, V, n, or T whenever three of these four are known. The Ideal Gas Law is a good approximation for real gases at low pressures (less than a few atmospheres) and high temperatures (much greater than the boiling point of the liquid phase).

The Ideal Gas Law assumes that atoms and molecules are points with no volume, and that they do not interact except through collisions. These assumptions are good when the density is low so the atoms and molecules are far apart, and the temperature is high so the interaction energy is small compared to the kinetic energy of the atoms and molecules.

When you look at Equation (1), you should have several questions about it. The *Exploration* below will help you address these questions, understand what the equation is telling you, and identify how you can use it.

Exploration

1. When used with the Ideal Gas Law, the gas constant R has units of L·atm·mol^{-1}·K^{-1}. Consequently, what units must be used for volume, pressure, amount of substance, and temperature in the Ideal Gas Law?

2. What is the numerical value for the gas constant, R? Using the following information obtained from experimental measurements on many gases, calculate a numerical value for the gas constant R. At 273 K and 1 atm pressure, 1 mole of these gases occupied 22.4 L. A temperature of 273 K and a pressure of 1 atm are called the standard temperature and pressure for gases (STP). The volume of 22.4 L is called the standard molar volume of an ideal gas.

3. What predictions can be made from the Ideal Gas Law, when the following changes occur?
   (a) The Kelvin temperature doubles while n and V remain unchanged.

   (b) The volume doubles while n and T remain unchanged.

   (c) Gas is added to a container, while T and P remain unchanged.

Application

1. A valve is opened between a 10 L tank containing a gas at 5 atm of pressure, and an empty 15 L tank. What is the new pressure of the gas, assuming the temperature does not change?

2. In Havre, Montana the temperature went from $-34°F$ on a day in January to $103°F$ on a day in July in 2005. If you inflated your automobile tires to a pressure of 30 lb/in² on that day in January, what would the pressure be on that day in July, using lb/in²? Assume the volume of the tire and the amount of air in the tire did not change.

3. A 1.00 g sample of water vaporizes completely inside a 10.0 L container at 150°C. What is the pressure of the water vapor?

4. A weather balloon is filled with helium to a volume of 1000. L on the ground where the pressure is 745 torr and the temperature is 20°C. When the balloon ascends to a height of 2 miles, the pressure is 600. torr and the temperature is $-33°C$. What is the volume of the helium in the balloon at this altitude?

Got It!

1. You've *Got It!* if you can answer this question without doing any numerical calculations. Which one of the following gas samples contains the largest number of molecules, and which one contains the smallest? Explain your reasoning.
   (a) 1.0 L of H_2 a STP
   (b) 1.0 L of N_2 at STP
   (c) 1.0 L of H_2 at 27°C and 760 torr.
   (d) 1.0 L of CO_2 at 0°C and 800 torr.

2. If a gas in a jar with a leaky seal (P and V constant) is heated and the Kelvin temperature quadruples, what fraction of the gas remains in the jar, and what fraction of the gas escapes?

3. Reconsider the predictions that you made at the beginning of this activity. If you decide to change any of your predictions in view of what you learned from this activity, identify which ones and explain why.

10-2. Partial Pressures: What determines the total pressure of a mixture of gases?

What do you think?

Suppose you have two 1.0 L containers at 300 K. One contains nitrogen gas, and the other contains oxygen gas. The pressure inside both containers is 1.0 atm. You transfer the contents of these two containers into a third, also 1.0 L at 300 K, as shown in the diagram below. Under these conditions, nitrogen and oxygen behave as ideal gases.

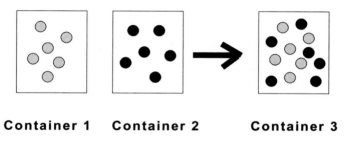

Container 1   Container 2   Container 3

Fig. 1. Oxygen and nitrogen gases from separate containers are combined into one container. The containers are identical.

1. Has the pressure produced by oxygen changed?

2. Has the pressure produced by nitrogen changed?

3. What is the total pressure inside Container 3?

Exploration

1. For the situation described by Figure 1, determine the following items using the Ideal Gas Law.
   (a) The number of moles of oxygen in Container 1, and the number of moles of nitrogen in Container 2.

   (b) The total number of moles in Container 3.

(c) Show how you can calculate the total pressure of the gases in Container 3 by using the Ideal Gas Law.

(d) Show how you can calculate the pressure of oxygen in Container 3 by using the ideal gas law with n_{O2} = moles of oxygen. This pressure is called the *partial pressure* of oxygen.

(e) Show how you can calculate the pressure of nitrogen in Container 3 by using the ideal gas law with n_{N2} = moles of nitrogen. This pressure is called the *partial pressure* of nitrogen.

2. Use your results from Questions 1 (c) and (d) to calculate the ratio of the partial pressure of oxygen to the total pressure.

3. Use your result from Question 1 (a) and (b) to calculate the ratio of the moles of oxygen to the total number of moles. This ratio is called the *mole fraction* of oxygen present in the mixture.

4. Are the values you calculated in Questions 2 and 3 equal or different? Divide the ideal gas law equation for the partial pressure of oxygen ($P_{O2}V = n_{O2}RT$) by the ideal gas law equation for the total pressure to show why the mole fraction of a gas must equal the ratio of partial pressure of that gas to the total pressure.

Information

John Dalton was the first to observe that the total pressure exerted by a mixture of gases is the sum of the partial pressures of the individual gases in the mixture. This statement is known as Dalton's law of partial pressures, which can be written in the form of an equation.

$$P_{total} = \sum_{sum\ over\ all\ gases} P_i \tag{1}$$

where P_i is the partial pressure of gas i.

This property is a consequence of the fact that ideal gas molecules act independently so the pressure of a gas inside a container depends on the number of molecules but not their identity. The pressure of one mole of oxygen plus one mole of nitrogen therefore will be the same as the pressure of two moles of nitrogen or two moles of oxygen.

This idea follows from the Ideal Gas Law as shown below.

For a mixture of 3 gases

$$P_{total} = n_{total} \frac{RT}{V}$$

$$n_{total} = n_1 + n_2 + n_3$$

Substituting for n gives

$$P_{total} = (n_1 + n_2 + n_3) \frac{RT}{V}$$

$$P_{total} = \left(n_1 \frac{RT}{V} + n_2 \frac{RT}{V} + n_3 \frac{RT}{V} \right)$$

$$P_{total} = P_1 + P_2 + P_3$$

Complete this proof by dividing $P_1 = n_1 RT/V$ by $P_{total} = (n_1+n_2+n_3)(RT/V)$ to obtain $P_1/P_{total} = n_1/n_{total}$ since the RT/V factors cancel.

So the partial pressure of a gas equals its mole fraction in a mixture times the total pressure of all the gases.

$$P_i = X_i P_{total}, \text{ where } X_i = \frac{n_i}{n_{total}} \tag{2}$$

Got It!

Mixtures of 15% oxygen and 85% helium (in terms of mole fraction) are used in some scuba diving tanks to help prevent the bends, which is a condition caused by nitrogen bubbles forming in tissues and blood. A standard scuba tank has a volume of 18 L and can be filled to a pressure of 3000 psi. Assume the tanks are filled on a hot day when the temperature is 30°C. (14.7 psi = 1.00 atm)

1. What are the partial pressures of oxygen and helium in one of these tanks?

2. How many liters of oxygen at STP need to be used to fill one of these tanks?

3. How many liters of helium at STP need to be used to fill one of these tanks?

4. How many moles of oxygen and moles of helium are used in filling one of these tanks?

5. What is the total mass of the gases in one of these tanks when it is filled?

108

11-1. Phases of Matter: As water cools, does the temperature continue to drop as ice forms?

What do you think?

1. You have ice-cold Coke with ice in a glass ice sitting on a table on a very hot day. Is the temperature of the Coke increasing, decreasing, or staying the same as the ice melts?

2. You have water in a pot boiling on the stove. Is the temperature of the water increasing, decreasing, or staying the same?

3. When water boils, what is inside the bubbles that form?

Information

There are three states of matter (solid, liquid, and gas), each with distinct characteristics. A *state* of matter also is called a *phase*. The solid state of a substance, however, can have more than one phase. Each solid state phase has a different structure. For example, diamond and graphite are two solid state phases of carbon.

Diamond consists of a three dimensional network of covalently bonded carbon atoms. In diamond, each carbon atom is bonded to four others, and the atomic orbitals are sp^3 hybridized.

Graphite has layers of a two dimensional network of covalently bonded carbon atoms. In graphite, each carbon atom is bonded to three others, and the atomic orbitals are sp^2 hybridized.

Transitions between one phase and another can be caused by heating or cooling to change the temperature or by increasing or decreasing the pressure to change the spacing between the molecules (or atoms for substance not composed of molecules like diamond, graphite, and the noble gases)

109

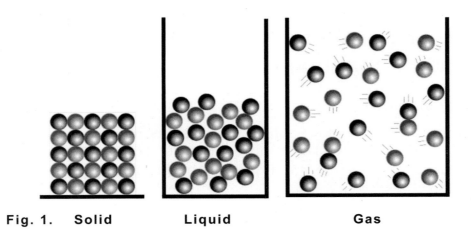

Fig. 1.   Solid   Liquid   Gas

Table I. Characteristics of the States of Matter

Solid	Liquid	Gas
Long range order	Short range order	Disordered
Tightly packed	Closely spaced	Widely spaced
Vibrate about fixed position	Limited motion – rotate and diffuse	Free motion
Almost incompressible	Not very compressible	Highly compressible
Fixed shape	No fixed shape	No fixed shape
Does not flow	Can flow	Can flow

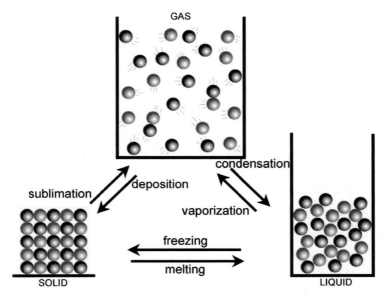

Fig. 2. The three states of matter and the six phase transitions between them.

Heating Curve for Water

amount of heat added

Fig. 3. Heating curve for water at 1 atm pressure showing how the temperature changes as heat is added at a constant rate.

Exploration

1. In view of Figure 1 and the information in Table I, what do you see as the three most significant differences among the three states of matter?

2. What are the three processes listed in Figure 2 that convert one phase into another when a substance is heated?

3. What are the three processes listed in Figure 2 that convert one phase into another when a substance is cooled?

4. Figure 3 shows a heating curve for water. This data is obtained by measuring the temperature when heating at a constant rate. Five regions are identified in the figure and are labeled A through E. Below 0°C, water is in the solid state (ice), between 0 and 100°C water is in the liquid state, and above 100°C water is in the gas state (steam).

(a) Next to the labels A, C, and E in Figure 3, write the state of water (solid, liquid, gas) in that region of the graph.

(b) What is happening to the water at the temperature labeled B in the figure?

(c) What is happening to the water at the temperature labeled D in the figure?

(d) Why does the temperature start to rise again at the points labeled B and D in the figure?

5. Based on the graph in Figure 3, do you think it takes more energy to melt 100 g of ice, raise the temperature of 100 g of water from 0°C to 100°C, or to boil and vaporize 100 g of liquid water? Explain.

Information

When energy is added to a substance, it increases the kinetic energy of the molecules and hence the temperature. At a phase transition, the energy is used to overcome attractive forces between the molecules, so the kinetic energy and temperature do not change.

Got It!

1. How can heating curves be used to identify the temperatures at which phase transitions occur for any given substance at any fixed pressure?

2. Reconsider your *What do you think?* responses at the beginning of this activity. If you decide to change any in view of what you have learned, identify which ones and explain why.

11-2. Phase Diagrams: What information can be obtained from a phase diagram?

A plot of the temperatures and pressures at which phase transitions occur is called a *phase diagram*. A generic phase diagram is shown below.

Fig. 1. A generic phase diagram showing the regions of temperature and pressure where the substance is a solid, liquid, and gas.

Exploration

1. In a phase diagram, areas represent different phases, and the boundary lines show the temperatures and pressures where two phases are in equilibrium and a phase transition occurs. (a) What is the boundary line called in Figure 1 where the liquid and gas phases are in equilibrium? What two phase transitions occur along this boundary line?

(b) What is the boundary line called in Figure 1 where the solid and liquid phases are in equilibrium? What two phase transitions occur along this boundary line?

(c) What label would you propose for the boundary line where the solid and the gas phases are in equilibrium? Add your label to Figure 1. What two phase transitions occur along this boundary line?

2. What phases are present in equilibrium at the triple point?

3. Along the vapor pressure curve, both the liquid and gas phases are present. If you look at a sample in a glass tube, you can see both these phases, just like looking at water in a bottle. You see the liquid at the bottom, and the water vapor (gas) is present in the space above it. As you increase the temperature, the vapor pressure increases. As you raise the temperature, what do you expect to see at the critical point?

Application

Phase diagrams for water and carbon dioxide are given below. Use these phase diagrams to answer the following questions about the properties of water and carbon dioxide. Note that the pressure scale for water is in mm Hg (torr), and the pressure scale for carbon dioxide is in atmospheres.

Fig. 5. Phase diagrams for water, H_2O (left),
and carbon dioxide, CO_2 (right).

1. Use the phase diagram to estimate the vapor pressure of water at 50°C.

2. Will ice sublime at some pressure and temperature? If not, why not? If so, below what temperature and pressure will ice sublime?

3. Label the triple point for carbon dioxide. What does this point represent?

4. (a) What is the lowest pressure and temperature at which liquid carbon dioxide is stable?

   (b) Identify the stable phase of carbon dioxide at temperatures below this point.

   (c) Identify the stable phase of carbon dioxide at pressures below this point.

5. Does the melting point of water increase or decrease with increasing pressure?

6. Does the melting point of carbon dioxide increase or decrease with increasing pressure?

7. The density of ice is less than the density of liquid water. Do you think increasing the pressure on ice facilitates or hinders melting? Explain and use your explanation to explain how pressure affects the melting point of ice.

8. The density of solid carbon dioxide is greater than the density of liquid carbon dioxide. Do you think increasing the pressure on solid carbon dioxide facilitates or hinders melting? Explain and use your explanation to explain how pressure affects the melting point of carbon dioxide.

Research

Use your textbook as a resource and explain why, in terms of noncovalent intermolecular interactions, the triple point of carbon dioxide occurs at a higher pressure and a lower temperature than the triple point of water.

Got It!

1. How many phases are in equilibrium along a boundary line?

2. How can you identify the normal boiling point of a substance on a phase diagram if it isn't labeled?

3. How can you identify the normal melting point of a substance on a phase diagram if it isn't labeled?

4. How can you identify the critical point of a substance on a phase diagram if it isn't labeled?

5. Sketch a heating curve obtained for carbon dioxide at a pressure of 10 atm. Label each phase and the phase transitions on your heating curve.

12-1. Solubility: Where do those vitamins go?

What do you think?

Dietary supplements, such as vitamins, are a multi-billion dollar industry. Over 40% of the adult population takes dietary supplements with vitamins C and E being among the most popular. Identify three things you should know about vitamins before deciding whether or not to take them as a dietary supplement.

Information

Vitamins are small molecules that are essential for our health. Most of the vitamins that we need are obtained from food and are not synthesized in our bodies. Vitamins are divided into two categories: those that are water soluble (*hydrophilic*) and those that are not water soluble (*hydrophobic*). The hydrophilic vitamins dissolve in the blood. The hydrophobic vitamins do not dissolve in the blood directly but are transported by blood lipoproteins and are absorbed by fatty membranes and tissues.

Some people advocate taking mega doses of vitamins to improve health, but large doses of the hydrophobic vitamins can cause harmful effects because they are stored with fat in cells and dissolve in fatty tissues, including cell membranes. Consequently, a significant concentration is maintained in the bloodstream, and concentrations in tissues build up over time. The hydrophilic vitamins, on the other hand, are not soluble in the fatty tissues; they remain dissolved in the blood and are eliminated quite rapidly through the kidneys. Consequently, they do not build up high concentrations in the body.

Vitamin C is an antioxidant and a coenzyme. It is known that it prevents scurvy. As an antioxidant, it neutralizes reactive molecules in cells (free radicals), reduces damage to the cells, and thereby prevents cancer. As a coenzyme, it contributes to the enzymatic synthesis of other essential molecules, such as collagen, which is the main connective protein in joints.

Vitamin E actually is a group of 8 isomers with structures similar to the one shown in Figure 1. A number of studies indicate that vitamin E, which functions as a powerful antioxidant, may prevent heart disease, cancer, cataracts, Alzheimer's disease, and age-related macular degeneration.

The structures of vitamin C and vitamin E are shown below.

Fig. 1.   Vitamin C (ascorbic acid)   Vitamin E

Exploration

1. Using the structures in Figure 1, identify the types of intermolecular interactions that are possible for vitamin C and for vitamin E. (Consider London dispersion, dipole-dipole, hydrogen bonding, and ionic. Use your textbook as a resource if you have not yet learned about these interactions.)

   Vitamin C:

   Vitamin E:

2. Which of these intermolecular interactions do you expect to be most important for vitamin C? Explain.

3. Which of these intermolecular interactions do you expect to be most important for vitamin E? Explain.

118

4. Considering your answers to *Exploration* questions 2 and 3 above, which vitamin do you expect to be *hydrophilic* and which one do you expect to be *hydrophobic*? Explain.

5. Considering your answers to *Exploration* questions 2 - 4 above and the structures shown in Figure 2 below, which vitamin, C or E, do you expect to be absorbed better by fatty tissue and scavenge free radicals in phospholipid cell membranes? Explain.

Fig.2.   triglyceride fat

phospholipid cell membrane

6. Based on the discussion in the *Information* section and your identification of the vitamins C and E as either *hydrophilic* or *hydrophobic*, which one do you think will remain in the bloodstream longer? Explain.

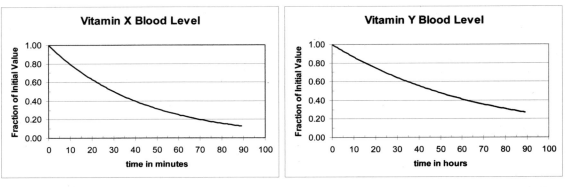

Fig. 3. Graphs showing the decrease in concentration of two vitamins, labeled X and Y, in the blood as a function of time after the initial dose. Note the difference in the x-axis scale, minutes for X and hours for Y.

7. Replace X and Y in the titles of the graphs in Figure 3 with the name of the corresponding vitamin, C and E, as appropriate. Explain your choice.

8. From the graphs in Fig. 3, determine the half-life of each vitamin in the bloodstream. The half-life is the time it takes for the concentration to drop to one-half of its initial value.

9. How long will it take for each vitamin concentration to drop to 10% of its initial value in the bloodstream?

10. How often do you think you need to take each vitamin to keep the concentrations above 50% of the initial concentration? Explain.

11. Draw concentration *vs* time graphs showing how repeated doses of each vitamin would keep the concentration between 50% and 150% of the initial peak concentration.

12. Based your analysis, what is your recommendation for how often you should take Vitamin C and E to maintain a consistent level of each vitamin in your bloodstream? Explain.

Got It!

1. Predict which one of the following vitamins would be retained in the bloodstream longer. Explain.

Vitamin B3, niacin   **Vitamin D**

2. Amoxicillin and azythromicin are common antibiotics. Amoxicillin usually is taken every 6 hours for 10 days, while azythromicin is taken once a day for 5 days. Based on these regimens, what can you predict about their half-lives and molecular structures? Explain.

Acknowledgement

This activity was adapted from a draft titled, *Vitamin C and E: How long do I have to stay here?* that was produced by Renee Cole, Linda Hobart, and Fred Yost at the POGIL-IC workshop, Litchfield, South Carolina, January, 2007.

12-2. Colligative Properties: What effect does a solute have on properties of a solvent?

What do you think?

If you add salt to water, does the water
(a) boil at a higher or lower temperature?

(b) freeze at a higher or lower temperature?

Information

A *solution* is a homogeneous mixture of two or more pure substances. The *solvent* is the component present in the largest amount. An aqueous solution has water as the solvent. The *solute* is the substance present in the smaller amount that is dissolved in the solvent. The *concentration* of a solution specifies the amount of solute present in some amount of the solution or solvent. Two common concentration units are molarity (M), which specifies the moles of solute per liter of solution, and molality (m), which specifies the moles of solute per kilogram of solvent.

Colligative properties are properties of dilute solutions that depend only on the number of nonvolatile solute particles and not on the types of particles present. Colligative properties include a lowering of the vapor pressure, elevation of the boiling point, depression of the freezing point, and osmotic pressure. All these effects are caused by adding a nonvolatile solute to a solvent, which reduces the vapor pressure of the solvent.

Since in a given amount of solution, some solvent molecules have been replaced by nonvolatile solute particles, fewer solvent molecules will be able to reach the surface and escape. Consequently the vapor pressure of the solution is smaller.

The boiling point is the temperature at which the vapor pressure of the solvent equals 1 atm. Since the solute has lowered the vapor pressure, then the boiling point of the solution necessarily will be higher because a higher temperature is needed to produce a vapor pressure of 1 atm.

Depression of the freezing point and osmotic pressure also can be traced to a decrease in vapor pressure caused by the nonvolatile solute.

The molar concentration of a solution varies with temperature because the volume of a solution is temperature dependent. Since some colligative properties involve temperature changes, molal units, which are temperature independent, are used in describing them.

Table I. Colligative Properties of Solutions

Property	Equation	Symbols
Vapor Pressure	$P = XP^o$	P = vapor pressure of the solvent over the solution P^0 = vapor pressure of the pure solvent X = mole fraction of the solvent in the solution
Boiling Point Elevation	$\Delta T_b = K_b mi$	ΔT_b = change in boiling point K_b = solvent boiling point elevation constant m = molal concentration of the solute i = van't Hoff factor = the number of moles of particles produced by 1 mole of solute
Freezing Point Depression	$\Delta T_f = - K_f mi$	ΔT_f = change in freezing point K_f = solvent freezing point depression constant m = molal concentration of the solute i = van't Hoff factor = the number of moles of particles produced by 1 mole of solute
Osmotic Pressure	$\Pi = cRTi$	Π = osmotic pressure c = molar concentration of the solute R = gas constant T = temperature i = van't Hoff factor = the number of moles of particles produced by 1 mole of solute

Exploration

1. What is the distinguishing characteristic of a colligative property?

2. What are four colligative properties of dilute solutions?

3. What are three things that these colligative properties have in common that can be identified from Table I?

4. How can one use the formula in Table I for the osmotic pressure to determine the molar mass of a polymer by dissolving it in some solvent?

Application

Use the formulas in Table I to answer the following questions.

1. People living at high altitudes where the pressure is less than 1 atm, *e.g.* Denver, often add salt to water to increase the boiling temperature when cooking. Is this a significant effect? Check by determining the boiling point of 1.0 L of water when 1 tablespoon of NaCl is added to it. A tablespoon of salt is about 15 g.

2. People living in cold climates, *e.g.* Minnesota, often throw some salt on icy sidewalks to melt the ice. If $CaCl_2$ is used, what is the molal concentration of the resulting solution if the ice melts at $-10°C$?

3. Order the following aqueous solutions in order of increasing osmotic pressure:
0.10 M KCl, 0.10 M sucrose, 0.050 M K_2SO_4, and 0.10 M $CaCl_2$.
.

4. Calculate the average molar mass of polyethylene when 4.40 g of the polymer is dissolved in benzene to produce 200.0 mL of solution and the osmotic pressure is found to be 7.60 torr at 25°C.

5. Sulfur exists in many forms with the general molecular formula S_n. If 0.48 g of sulfur is added to 200 g of carbon tetrachloride, CCl_4, and freezing point is depressed by 0.28°C, what is the molar mass and molecular formula of the sulfur?

Got It!

1. Two beakers, one (labeled A) containing 100.0 mL of a 1.0 M salt solution (NaCl), and the other (labeled B) containing 100.0 mL of pure water are placed inside a larger container, which then is sealed.
   (a) What do you think happens to the liquid levels in the two beakers as time elapses? Explain your reasoning.

   (b) Sketch a graph showing how the concentration of the solution in Beaker A changes with time.

2. (a) In what ways is the experiment in *Got It* question 1 similar to an osmosis experiment where two compartments of a container are separated by a membrane that only allows solvent molecules to pass. On compartment is filled with pure solvent, and the other compartment is filled with a salt solution.

   (b) What happens to the liquid levels in those two compartments as time elapses?

13-1. Rates of Chemical Reactions: How quickly do reactants turn into products?

The *rate* of any process is the change in some quantity over some period of time. For example, if you walk 1 mile over a period of 20 minutes, the rate at which your position changes is 1 mile/20 min = 0.05 mile/min, which is called your speed. Usually such speeds are expressed in miles per hour, in these units your walking speed is (0.05 mile/min)(60 min/hr) = 3 mile/hr.

What do you think?

1. Identify 3 examples of something that you have observed changing over a period of time.

2. For each of the examples that you identified in #1 above, estimate the rate for that process.

Exploration

Peroxyacetylnitrate (PAN) is a powerful respiratory and eye irritant found in smog. It is produced from pollutants (unburned hydrocarbons, NO, and NO_2) in the exhaust of power plants and internal combustion engines.

In an undergraduate research project, a team of students measured the rate of decomposition of PAN. Their results are given in Table I. The initial concentration of PAN was 50.0×10^{-6} M, and Table I shows how the concentration decreased with time. The concentration was monitored over a period of 91 minutes. The beginning of the experiment is designated as t = 0 (t = time).

127

Table I. Decomposition of Peroxyacylnitrate (PAN).

time (min)	conc (10^{-6} M)	\<rate\>	\<rate\>/conc
0	50.0		
1	48.3	1.740	0.0348
10	35.1		
11	33.9	1.221	0.0348
20	24.6		
21	23.8		
30	17.3		
31	16.7		
40	12.1		
41	11.7		
50	8.5		
51	8.2		
60	6.0		
61	5.8		
70	4.2		
71	4.0		
80	2.9		
81	2.8		
90	2.1		
91	2.0		

The students began to analyze the data by calculating the average rate of the decomposition reaction over a 1 minute interval after every 10 minutes by using Equation (1). The results of these calculations are in the third column of the table.

$$Rate = -\frac{\Delta[PAN]}{\Delta t} = -\frac{[PAN]_{t_2} - [PAN]_{t_1}}{t_2 - t_1} \tag{1}$$

They also divided the average rate by the concentration at the beginning of each 1 minute interval. Those numbers are in the fourth column of the table.

1. The students ran out of time to complete their analysis. Please help them by providing the missing entries in the third and fourth columns of Table I. Enter the values that you calculate in the unshaded rectangles in the table.

2. Show how the concentration of PAN changes with time by making a graph in the space below. Plot the concentration of PAN, which is in the second column of the table, on the y-axis and time on the x-axis. Be sure to label the axes, provide a scale for each axis, and include a title for the graph.

3. The slope of your graph at any point indicates the instantaneous reaction rate at that time. Draw lines on your graph to indicate the initial rate, the rate at 30 s, and the rate at 90 s.

4. Does the rate of the reaction increase, decrease, or remain the same as the concentration of PAN decreases? How can you tell from your graph?

5. How long did it take for half the PAN to decompose? This time is designated as the half-lifetime or the *half-life* of the decomposition reaction. Draw a vertical line on your graph to mark the half-life time.

6. What happens to the ratio of the reaction rate to the concentration of PAN as the reaction progresses? This information is given in the fourth column of Table I.

7. In view of your answers to *Exploration* questions 1 – 6, which equation below, where k = a constant, describes the data in Table I? Explain or justify your choice.

   a) Rate = k

   b) Rate = k [PAN]

   c) Rate = k [PAN]^2

Information

A rate law for a reaction tells how the rate of the reaction depends on the concentrations of the chemical species involved in the reaction. Rate laws have the form given by Equation (2). Equation (2) says that the rate of the reaction equals some constant multiplied by the product of concentrations [A], [B], and [C] of the reactants with exponents x, y, and z, respectively.

$$\text{Rate} = k[A]^x[B]^y[C]^z \qquad (2)$$

An exponent in the rate law gives the order of the reaction with respect to that chemical species. For example, if x and z equal 1 and y = 2, then the reaction is first order with respect to

A and C and second order with respect to B. The overall order of the reaction is the sum of the exponents in the rate law, which in this example would be x + y + z = 1 + 2 + 1 = 4.

The choices in *Exploration* question 7 correspond to the rate laws for a *zero-order* reaction, a *first-order* or unimolecular reaction, and a *second-order* or bimolecular reaction. The rate of a zero-order reaction does not depend on the concentration of the reactants, the rate of a first-reaction depends linearly on the concentration, and the rate of a second-order reaction depends on the square of the concentration.

The order of a reaction often differs from the stoichiometric coefficients in the balanced reaction equation. The stoichiometric coefficients relate the numbers of reactant and product molecules involved in the reaction. The order of the reaction depends on the reaction mechanism, not on the number of molecules involved in the overall reaction.

The rate of a reaction always is expressed as a positive number and can be written in terms of the change in concentrations of any of the reactants or products. For example, for the reaction $2H_2 + O_2 \rightarrow 2H_2O$, the rate could be written in any of the following ways.

$$\text{a)} \quad Rate = -\frac{1}{2}\frac{\Delta[H_2]}{\Delta t}$$

$$\text{b)} \quad Rate = -\frac{\Delta[O_2]}{\Delta t} \tag{3}$$

$$\text{c)} \quad Rate = \frac{1}{2}\frac{\Delta[H_2O]}{\Delta t}$$

In (3a) and (3b) minus signs are needed to make the rate a positive number because the concentration is decreasing, so the change in concentration is negative. In (3a) and (3c) the change in concentration needs to be multiplied by ½ because those concentrations are changing twice as fast as the oxygen concentration and indicated by the stoichiometric coefficients in the balanced reaction equation.

Application

1. Is the decomposition of PAN a zero, first, or second order reaction? Explain.

2. Using the data in Table I, calculate the value and determine the units of the rate constant, k, for the decomposition of PAN.

3. Using the relationship $t_{1/2}$ = ln(2)/k, calculate the half-life, $t_{1/2}$, of PAN from the value you calculated for k in *Application* question 2. Is this calculated value consistent with your estimate of the half-life in *Exploration* question 5?

Information

The graph you made in the *Exploration* section has the form of an exponential function. This exponential function gives the concentration of PAN as a function of time.

$$[PAN(t)] = [PAN(t=0)] \, e^{-kt} \qquad (4)$$

Equation (4) is called the *integrated rate law* for a first order reaction. It is called the integrated rate law because its derivative gives the first order rate law, which you discovered in response to *Exploration* question 7.

$$Rate = -\frac{\Delta[PAN(t)]}{\Delta t} = k[PAN(t)] \qquad (5)$$

The integrated rate law, Equation (4), can be put in logarithmic form by taking the natural logarithm of both sides.

$$\ln([PAN(t)]) = -kt + [PAN(t=0)] \qquad (6)$$

Got It!

1. In view of Equation 6, what would you need to plot to produce a graph from the data in columns 1 and 2 of Table I with all the points lying on a straight line? Note that the equation for a straight line is y = mx + b.

2. Nickel is purified for use in making steel alloys by heating $Ni(CO)_4$ in a vacuum. It decomposes in a first-order unimolecular reaction: $Ni(CO)_4 \rightarrow Ni + 4CO$. At 66.0°C, the rate constant for this reaction is 1.873 s^{-1}.
   (a) Write the rate law for this reaction.

   (b) Calculate the time it would take for half the nickel tetracarbonyl to decompose.

   (c) Write the integrated rate law for this reaction.

   (d) Use the integrated rate law to calculate the time it would take for 90% of the nickel tetracarbonyl to decompose.

Additional Practice Problems

1. A reaction has the following experimental rate law. If the concentration of X is doubled, and the concentration of Y is halved, by what factor does the reaction rate change? Explain.

$$rate = k[X]^2[Y]$$

2. In addition to PAN described in the *Exploration*, another irritant in smog is formaldehyde, which can be formed by the reaction of ozone with ethylene. In three different experiments, the undergraduate research team collected the following data for this reaction but did not have time to analyze the data. Again, please help them.

$$2C_2H_4(g) + 2O_3(g) \rightarrow 4CH_2O(g) + O_2$$

Initial Concentration O_3 (mol/L)	Initial Concentration C_2H (mol/L)	Initial Rate of Formation of CH_2O $(mol \cdot L^{-1} \cdot s^{-1})$
0.50×10^{-7}	1.0×10^{-8}	1.0×10^{-12}
1.5×10^{-7}	1.0×10^{-8}	3.0×10^{-12}
1.0×10^{-7}	2.0×10^{-8}	4.0×10^{-12}

(a) Using logic similar to that used in answering the previous question, determine the rate law for the reaction from the data in the table.

(b) Use the data in the table to calculate a value for the rate constant, k.

(c) What would the rate of the reaction be if the initial concentrations of the reactants were both 2.0×10^{-7} M?

13-2. Reaction Mechanisms: What determines the rate law for a reaction?

Many chemical reactions do not occur in one step; rather, they involve a series of steps. The steps correspond to a single molecule dissociating or rearranging in some way, or to two molecules colliding and reacting with each other. Reaction mechanisms rarely involve the simultaneous collision of three or more molecules because it is extremely unlikely for three or more molecules to arrive at the same point in space at the same time. Such an event is similar to three cars arriving from different directions and colliding at an intersection: collisions of two cars at an intersection occur much more often.

Each step in the overall reaction is called an *elementary reaction*, and a reaction equation can be written for it. The set of reaction equations for all the elementary reactions is called the *reaction mechanism.*

A rate law can be predicted for any reaction mechanism, and the predicted rate law can be compared with the one observed experimentally. If the predicted and experimental rate laws disagree, then the proposed reaction mechanism is not the correct one. If they agree, one can only say that the mechanism is consistent with the experimental rate law. Agreement does not prove that the proposed mechanism is the correct one because different mechanisms can predict the same rate law.

What do you think?

Based on the previous activity, *Rates of Chemical Reactions*,

1. What is the rate law for a unimolecular elementary reaction, e.g. A → B?

2. What is the rate law for a bimolecular elementary reaction, e.g. A + B → C?

Exploration 1: Predicting a Rate Law

Nitric oxide, NO, is present in the exhaust of power generating plants and internal combustion engines. It reacts with the oxygen in air to produce nitrogen dioxide, NO_2, which causes smog to have a brownish color.

$$2NO(g) + O_2(g) \rightarrow 2NO_2(g)$$

Two mechanisms that have been proposed for this reaction are given below.

Mechanism 1
Step 1: $NO + O_2 \rightleftarrows NO_3$ (fast, equilibrium established)
Step 2: $NO_3 + NO \rightarrow 2NO_2$ (slow)

Mechanism 2
Step 1: $NO + O_2 \rightarrow NO_2 + O$ (slow)
Step 2: $NO + O \rightarrow NO_2$ (fast)

1.1. Any valid mechanism must consist of a series of unimolecular or bimolecular elementary reaction steps that give the overall reaction equation when added together.
(a) Add the 2 steps for Mechanism 1 together. Do they give the overall reaction equation?

(b) Add the 2 steps for Mechanism 2 together. Do they give the overall reaction equation?

1.2. The rate of the overall reaction is limited by and equal to the rate of the slowest step in the mechanism.
(a) What is the rate law for the slow step in Mechanism 1?

(b) What is the rate law for the slow step in Mechanism 2?

136

1.3. To compare with experimental measurements, the rate law for a reaction must be expressed in terms of the concentrations of reactants and products and not include the concentrations of intermediates, which are not easily measured. Do the rate laws that you have written in response to *Exploration* question 2 include only the reactants NO_2 and O_2?

1.4. To eliminate an intermediate from the rate law, assume that the fast reaction is reversible and that the concentration of the intermediate reaches a steady state or equilibrium value. In the steady state, the intermediate is destroyed as fast as it is formed. The following equations describe this steady state condition for Step 1 in Mechanism 1.

formation rate = decomposition rate

$$k_1[NO][O_2] = k_{-1}[NO_3] \qquad (1)$$

Rearranging this equation gives

$$[NO_3] = (k_1/k_{-1})[NO][O_2] \qquad (2)$$

(a) Using Equation (2), rewrite your the rate law that you predicted in *Exploration* question (1.2a) for Mechanism 1 in terms of only the concentrations of NO and O_2.

(b) Write the rate laws that you predict for Mechanisms 1 and 2 in the table below.

Table I. Rate Laws Predicted for the Oxidation of Nitric Oxide

Mechanism 1	Rate =
Mechanism 2	Rate =

Exploration 2: Determining the Rate Law Experimentally

The relationship between the concentration of reactants and the rate of the reaction is given by the rate law for the reaction. One way to determine this relationship experimentally is to measure the initial rate of the reaction when different concentrations of reactants are used. Such data collected for the oxidation of nitric oxide are given in Table II.

The rate law for the oxidation of nitric oxide has the following form: Rate = k $[NO]^x[O_2]^y$. The objective is to use the data in Table II to determine the exponents x and y in the rate law. You did this before at the end of Activity 13-1.

Table II. Initial Reaction Rates for the Oxidation of Nitric Oxide

Experiment	[NO] 10^{-2} M	[O$_2$] 10^{-2} M	Rate 10^{-3} M s^{-1}
A	1	1	7
B	2	1	28
C	1	2	14

2.1. According to the data in Table II, when the concentration of nitric oxide was kept constant and the concentration of oxygen increased by a factor of 2, by what factor did the initial rate of the reaction increase?

2.2. In view of your answer to Question 2.1, what is the value of the exponent y in the rate law, Rate = k $[NO]^x[O_2]^y$?

2.3. According to the data in Table II, when the concentration of nitric oxide increased by a factor of 2 and the concentration of oxygen was kept constant, by what factor did the initial rate of the reaction increase?

2.4. In view of your answer to Question 2.3, what is the value of the exponent x in the rate law, Rate = k $[NO]^x[O_2]^y$?

2.5. Summarize your results by writing the rate law for the reaction.

Got It!

1. Based on *Exploration 2*, identify which reaction mechanism, 1 or 2, is consistent with the experimental rate law.

2. Calculate a value for the rate constant for the oxidation of nitric oxide from the data in Table II.

3. Which of the reactions below are
   (a) unimolecular elementary reactions.

   (b) bimolecular elementary reactions.

   (c) not elementary reactions.

    I. $CH_4(g) + 2O_2(g) \rightarrow CO_2(g) + 2H_2O(g)$

    II. $O_3(g) + O(g) \rightarrow 2O_2(g)$

    III. $Mg(s) + 2H_2O(l) \rightarrow H_2(g) + Mg(OH)_2(s)$

    IV. $O_3(g) \rightarrow O_2(g) + O(g)$

    V. $HCl(g) + H_2O(g) \rightarrow H_3O^+(g) + Cl^-(g)$

    VI. $I^-(g) + CH_3Cl(g) \rightarrow CH_3I(g) + Cl^-(g)$

    VII. $N_2O_4(g) \rightarrow 2NO_2(g)$

    VIII. $N_2(g) + 3H_2(g) \rightarrow 2NH_3(g)$

4. The reaction rate for the bromination of acetone depends on the pH of the solution even though hydronium ion does not appear in the overall reaction equation. The decrease in the concentration of bromine was measured for several different initial concentrations as described in the table below.

$$CH_3COCH_3(aq) + Br_2(aq) \rightarrow CH_3COCH_2Br(aq) + HBr(aq)$$

Initial [CH$_3$COCH$_3$] (M)	Initial [Br$_2$] (M)	Initial [H$_3$O$^+$] (M)	Δ[Br$_2$] (M/s)
0.30	0.05	0.05	5.7×10^{-5}
0.30	0.10	0.05	5.7×10^{-5}
0.30	0.05	0.10	12.0×10^{-5}
0.40	0.05	0.20	31.0×10^{-5}
0.40	0.05	0.50	7.6×10^{-5}

(a) What is the rate law for this reaction?

(b) What is the order of the reaction with respect to each reactant?

(c) What is the overall order of the reaction?

(d) What is the value of the rate constant k for this reaction?

(e) What is the rate of the reaction when all the reactant concentrations are 0.01 M?

14-1. Equilibrium Constant and Reaction Quotient: How can the direction of a reaction be identified?

What do you think?

Which one of the endings (a), (b), or (c) to the following sentence do you think is correct? The concentration of products in a chemical reaction stops increasing when (a) all the reactants have been converted into products, (b) the concentrations of reactants and products are equal, or (c) the products are converted back into reactants at the same rate as reactants are being converted to products.

Information

When chemicals are mixed together, they generally react to produce products. The concentrations of the reactants decrease and the concentrations of the products increase until dynamic equilibrium is established. When dynamic equilibrium is established, the concentrations of reactants and products no longer change because products are being converted back to reactants at the same rate as reactants are being converted into products.

This equilibrium condition is defined by an *equilibrium constant* for every reaction. For the following general reaction at equilibrium,

$$aA + bB \leftrightarrows cC + dD \tag{1}$$

the *equilibrium constant expression* is written as,

$$K_c = \frac{[C]_{eq}^c [D]_{eq}^d}{[A]_{eq}^a [B]_{eq}^b}. \tag{2}$$

The equilibrium constant is a ratio of the molar concentrations of the products divided the molar concentrations of the reactants with each concentration raised to the power given by the corresponding stoichiometric coefficient in the balanced reaction equation. The numerical value of this ratio is called the equilibrium constant.

Up to now, you probably have implicitly assumed that reactions go to completion, i.e. that reactants are completely converted into products unless one reactant is limiting. This situation of complete conversion is quite common and means that the equilibrium constant calculated from equation (2) is very large

141

because the concentrations of the reactants, A and B, are very small, essentially zero. However, other situations are common too, and equilibrium constants can have any value.

The *reaction quotient* has the same mathematical form as the equilibrium constant expression except it involves the actual concentrations in a reaction mixture instead of the equilibrium concentrations.

$$Q = \frac{[C]^c [D]^d}{[A]^a [B]^b} \qquad (3)$$

Exploration

1. What items go into the numerator of the equilibrium constant expression?

2. What items go into the denominator of the equilibrium constant expression?

3. What determines the exponent for each concentration in the equilibrium constant expression?

4. What are three similarities in the expressions for the equilibrium constant and the reaction quotient?

5. What do you see as the most important difference between the expressions for the equilibrium constant and the reaction quotient?

6. Would you expect a reaction with a large value for the equilibrium constant to be reactant favored or product favored? Explain.

7. If Q is larger than K (Q > K) for a given reaction mixture, would you expect the reaction to proceed in the forward direction to produce more products or in the reverse direction to produce more reactants? Explain.

8. For a given reaction mixture, when will a value calculated for Q equal the value for K found in a table of equilibrium constants?

Application

1. In an automobile engine, nitrogen and oxygen combine to form nitric oxide. Write the equilibrium constant expression, K_c, for this reaction.

$$N_2(g) + O_2(g) \leftrightarrows 2\ NO(g)$$

2. Calcium ions react with carbonate ions in water to form a solid deposit of calcium carbonate around water faucets. Write the equilibrium constant expression for this reaction. Note that the concentration of a pure solid (mol/L) does not change during a reaction; consequently, the concentration of the solid is not included in the equilibrium constant expression.

$$Ca^{2+}(aq) + CO_3^{2-}(aq) \leftrightarrows CaCO_3(s)$$

3. Acetic acid is an ingredient in vinegar. It reacts with water to produce hydronium ions and acetate ions. Write the equilibrium constant expression for this reaction. Note that the concentration of water is very large (55.5 mol/L) and does not change significantly as a result of the reaction, so the concentration of water or any other pure liquid is not included in the equilibrium constant expression.

$$CH_3COOH(aq) + H_2O(l) \leftrightarrows H_3O^+(aq) + CH_3COO^-(aq)$$

Got It!

1. Explain how the value of the equilibrium constant can help you decide whether or not products predominate over reactants when equilibrium has been achieved for any reaction.

2. Given the concentrations of reactants and products, explain how the values of the equilibrium constant and the reaction quotient can help you identify the direction of the reaction, i.e. whether the reaction will go forward to produce more products or reverse to produce more reactants.

3. The equilibrium constant for the reaction of hydrogen, H_2, with iodine, I_2, to form hydrogen iodide, HI, has a value of 50.0 at 745 K.
   (a) What is the value of the reaction quotient for a mixture of 0.5 mol H_2, 0.5 mol I_2, and 5.0 mol HI in a 2 L container at 745 K?

   (b) Will more HI be formed or will HI decompose as the reaction mixture approaches equilibrium? Explain.

144

14-2. Calculating Equilibrium Concentrations: How can the equilibrium concentrations of reactants and products be determined?

In order to make coal a cleaner and more transportable fuel, methods are being developed to produce combustible gases from it. In one of these methods carbon monoxide produced by burning coal with a limited amount of oxygen is reacted with steam to produce hydrogen.

$$CO(g) + H_2O(g) \rightleftarrows CO_2(g) + H_2(g)$$

Researchers in this area need to know how much carbon monoxide and water can be converted into hydrogen under various conditions. For example, if 1 mol each of carbon monoxide and steam were placed in a 20.0 L container at 420°C, what would be the concentrations of the four gases when equilibrium is reached? The equilibrium constant for this reaction is 10 at 420°C.

Such equilibrium problems are easily solved if you use a systematic procedure. One such procedure involves constructing a RICE table, as shown below. The letters RICE refer to the rows in the table and what you need to do to solve the problem: R = write the balanced <u>reaction equation</u>, I = specify the <u>initial amounts</u> that you are given, C = identify the <u>change</u> caused by the reaction, and E = identify the <u>equilibrium concentrations</u> in terms of the initial amounts and the change.

Table I. RICE Table for the Reaction of CO and H₂O

R	CO(g)	H₂O(g)	⇆	CO₂(g)	H₂(g)	Write the balanced reaction equation.
I	0.05	0.05		0	0	Initial concentrations (mol/L)
C	-x	-x		+x	+x	Change caused by the reaction (mol/L)
E	0.05-x	0.05-x		+x	+x	Concentrations at equilibrium (mol/L)

To determine the concentrations at equilibrium (row 4), an unknown amount (x in row 3) is assumed to react to reach equilibrium. The concentrations at equilibrium (row 4) then are substituted into the equilibrium constant expression and the resulting equation is solved to find a value for x, the amount that reacts, as shown below.

145

$$K_c = \frac{[CO_2][H_2]}{[CO][H_2O]}$$

$$10 = \frac{x^2}{(0.05-x)^2}$$

Taking the square root of both sides produces

$$3.16 = \frac{x}{0.05 - x}$$

Rearranging and solving for x produces

$$3.16(0.05 - x) = x$$

$$x = 0.158/45.16$$

$$x = 0.0380$$

The equilibrium concentrations therefore are 0.012 M for both CO and H_2O (which is 0.05-x), and 0.038 M for both CO_2 and H_2(which is just x).

Exploration

1. Why is writing the balanced reaction equation an important part of the procedure for solving equilibrium problems?

2. How would the third and fourth lines in Table I be changed if the stoichiometric coefficients in the balanced reaction equation were different than 1?

3. Why is the last line in Table I an especially important one?

4. What insight about solving equilibrium problems have you gained by examining the above example and procedure?

146

Information

Generally substituting the equilibrium concentrations into the equilibrium constant expression will produce a quadratic equation for x. A quadratic equation has the following form, where a, b, and c are constants.

$$ax^2 + bx + c = 0$$

A quadratic equation has 2 possible solutions, which are given by the following equation. While both solutions are mathematically valid, generally only one of the solutions makes physical sense and is consistent with the situation.

$$x = \frac{-b \pm \sqrt{b^2 - 4ac}}{2a}$$

In *Application* 2 below, K_c is small, which means that the amount that reacts, x, is small. You then can solve the equation by assuming that x is negligible, and $[NOCl] - x \approx [NOCl]$. You generally can make this assumption if the equilibrium constant is less than 10^{-3}, and if it turns out that the amount that reacts divided by the initial concentration is less than 0.05, which is the 5% convention.

Application

1. Carbon dioxide reacts with carbon to give carbon monoxide.

$$C(s) + CO_2(g) \leftrightarrows 2\ CO(g)$$

At 700°C, a 2.00 L flask contains an equilibrium mixture of 0.10 mol CO(g), 0.20 mol CO_2(g), and 0.40 mol C(s). Write the equilibrium constant expression, substitute the values for the concentrations, and calculate a value for the equilibrium constant for this reaction, K_c at 700°C.

2. At a certain temperature the equilibrium constant for the decomposition of NOCl is $K_c = 2.0 \times 10^{-5}$. Construct a RICE tble and calculate values for the equilibrium concentrations of NO and Cl_2 when 1.00 mol of NOCl is placed in a 1.00 L container.

$$2NOCl(g) \leftrightarrows 2\,NO(g) + Cl_2(g)$$

Got It

1. Why was it not necessary to solve a quadratic equation to determine the equilibrium concentrations of CO, H_2O, CO_2, and H_2 in the example at the beginning of the activity?

2. Why is it often not necessary to solve a quadratic equation in equilibrium problems when the equilibrium constant is small?

3. What amount of hydrogen iodide, HI, will be produced at equilibrium when 1 mol of iodine, I_2, and 3 mol of hydrogen, H_2, are mixed in a 10.0 L container at 745 K? $K_c = 50.0$ at 745 K for the following reaction.

$$H_2(g) + I_2(g) \rightleftharpoons 2\,HI(g)$$

Additional Practice Problems

1. Write the equilibrium constant expressions, K_c, for the following reactions.

$$CuCl_4^-(aq) \rightleftharpoons Cu^{2+}(aq) + 4Cl^-(aq)$$

$$LaCl_3(s) + H_2O(g) \rightleftharpoons LaClO(s) + 2HCl(g)$$

$$H_2S(g) + H_2O(l) \rightleftharpoons H_3O^+(aq) + HS^-(aq)$$

2. At equilibrium at 250°C, the following concentrations were found for the decomposition reaction: $PCl_5(g) \rightleftharpoons PCl_3(g) + Cl_2(g)$: $[PCl_5]= 4.2 \times 10^{-5}$ M, $[PCl_3] = 1.3 \times 10^{-2}$ M, and $[Cl_2] = 3.9 \times 10^{-3}$ M. What is the value of the equilibrium constant, K_c, for this reaction?

3. Nitrosyl chloride, NOCl, decomposes to nitric oxide, NO, and chlorine, Cl_2, at high temperatures. When 4.00 mol of NOCl was placed in a 2.00 L flask at a temperature of 462°C and some of it decomposes, the equilibrium concentration of NO was found to be 0.66 M. What is the value of the equilibrium constant, K_c, for this reaction?

4. Consider the reaction $N_2(g) + O_2(g) \rightleftarrows 2NO(g)$. At 2300 K the equilibrium constant $K_c = 1.7 \times 10^{-3}$. If 0.15 mol of nitric oxide, $NO(g)$, is placed in a 10.0 L flask and heated to 2300 K, what are the equilibrium concentrations of all three substances?

5. Consider the reaction $N_2(g) + O_2(g) \rightleftarrows 2NO(g)$. At 2300 K the equilibrium constant $K_c = 1.7 \times 10^{-3}$. Suppose 0.015 mol of $NO(g)$, 0.25 mol $N_2(g)$, and 0.25 mol $O_2(g)$ are placed in a 10.0 L flask and heated to 2300 K.
(a) Is this reaction at equilibrium initially? Explain.

(b) If not, in which direction must the reaction proceed to reach equilibrium? Explain.

(c) Calculate values for the equilibrium concentrations of all three substances.

15-1. Acid Ionization Constants: How strong is that acid?

Svante Arrhenius first defined an *acid* as a substance that produces hydronium ions (H_3O^+) when added to water, and a *base* as a substance that produces hydroxide ions (OH^-) when added to water.

Later Johannes Brønsted in Denmark and T. Martin Lowry in England independently proposed to give acids and bases more general definitions. According to the Brønsted-Lowry concept, an *acid* is a substance than donates a hydrogen ion or proton (H^+) to another substance, and a *base* is a substance that accepts a proton from another substance.

What do you think?

Identify some acids that you know are found in food products, *e.g.* in vinegar, citrus fruits, grapes, apples, and milk.

Information

In an aqueous solution, Brønsted-Lowry acids react with water and ionize. For example,

$$HF(aq) + H_2O(l) \rightleftharpoons H_3O^+(aq) + F^-(aq). \qquad (1)$$

In this reaction, water acts as a base and accepts a proton from a hydrofluoric acid molecule.

Such ionization reactions are described by an equilibrium constant, which is called the *acid ionization constant*, K_a. The acid ionization constant corresponding to the ionization of hydrofluoric acid in Equation (1) is defined as

$$K_a = \frac{\left[H_3O^+ \right]\left[F^- \right]}{\left[HF \right]} \qquad (2)$$

where the concentrations are the equilibrium concentrations. Values for some acid ionization constants are given in Table I.

Table I also contains pK_a values as defined by Equation (3). pK_a values are used because they are quicker to write and enter into a calculator than a K_a value.

$$pK_a = -\log(K_a). \qquad (3)$$

Table I. Acid ionization constants at 25°C

Acid	Formula	K_a	pK_a
Hydrochloric acid	HCl	1.0×10^4	-4.0
Sulfuric acid	H_2SO_4	1.0×10^3	-3.0
Nitric acid	HNO_3	2.0×10^1	
Phosphoric acid	H_3PO_4	7.5×10^{-3}	2.1
Hydrofluoric acid	HF	7.2×10^{-4}	
Nitrous acid	HNO_2	4.5×10^{-5}	4.3
Acetic acid	CH_3COOH		4.7
Carbonic acid	H_2CO_3	4.2×10^{-7}	6.4
Dihydrogen phosphate	$H_2PO_3^-$	6.2×10^{-8}	7.2
Ammonium ion	NH_4^+		9.3
Ethanol	CH_3CH_2OH	1.0×10^{-16}	16.0
Methane	CH_4	1.0×10^{-48}	48.0

Exploration

1. What is the definition of a Brønsted-Lowry acid?

2. What is the definition of a Brønsted-Lowry base?

3. Considering both the forward and reverse reactions in equation (1), what are the two Brønsted-Lowry acids in this reaction equation?

4. Considering both the forward and reverse reactions in equation (1), what are the two Brønsted-Lowry bases in this reaction equation?

5. What determined the order from top to bottom in which the acids are written in Table I?

6. Some K_a and pK_a values are missing in Table I. Use the ones that are given to calculate the ones that are missing by using equation (3), and complete the table.

7. For a 1.0 M solution of hydrofluoric acid, [HF] = 0.973 M, [H_3O^+] = 0.0264 M, and [F^-] = 0.0264 M. Show whether or not these concentrations are consistent with the acid ionization constant for hydrofluoric acid in Table I.

8. For a 1.0 M solution of hydrochloric acid, [HCl] = 0.0001 M, [H_3O^+] = 0.9999 M, and [Cl^-] = 0.9999 M. Show whether or not these concentrations are consistent with the acid ionization constant for hydrochloric acid in Table I.

9. To produce the concentrations in *Exploration* question 7, 0.0264 moles of the 1 mole of HF ionized. What percent of the HF molecules ionized?

10. To produce the concentrations in *Exploration* question 8, 0.9999 moles of the 1 mole of HCl ionized. What percent of the HCl molecules ionized?

11. Hydrofluoric acid is called a *weak acid* and hydrochloric acid is called a *strong acid*. Based on your results for *Exploration* questions 9 and 10, why do you think hydrofluoric acid is called a weak acid, and why do you think hydrochloric acid is called a strong acid?

Got It!

1. Based on your answer to *Exploration* question 11, write a definition for a *weak acid*.

2. Based on your answer to *Exploration* question 11, write a definition for a *strong acid*.

3. Using your definition of a strong acid, list all the strong acids that are given in Table I.

4. Using your definition of a weak acid, list all the weak acids that are given in Table I.

5. (a) Complete Table II by identifying each of the acids as strong or weak and calculating the K_a value.
   (b) Some weak acids are stronger than others, list the acids in Table II in order of decreasing strength.

Table II. Some acids found in foods.

Acid	Strong/Weak	Formula	pK$_a$	K$_a$
Lactic acid – found in milk products			3.88	
Oxalic acid – found leafy vegetables, like rhubarb and spinach			1.23	
Malic acid – found in sour or tart foods like green apples			3.40	

15-2. Calculations Involving Acid Ionization Constants: How can the pH of an acid solution be determined?

An acid ionization constant, K_a, is the equilibrium constant for an acid ionization reaction. It provides information about the concentrations of the reactants and products at equilibrium.

An important product in an acid ionization reaction is the hydronium ion, and the hydronium ion concentration often is expressed as a pH value using Equation (1).

$$pH = -\log([H_3O^+]) \tag{1}$$

What do you think?

In a previous activity (*Calculating Equilibrium Concentrations*) you learned how to connect the equilibrium constant to the equilibrium concentrations. The strategy involved constructing a *RICE Table*. Summarize the steps in this strategy below.

Exploration 1: Determining the pH of a Strong Acid Solution

What is the pH of a solution prepared by adding 0.25 mol of HCl to 125 mL of water? Solve this problem by completing the RICE table below.

(R) Write the balanced reaction equation for ionization of the acid.

$$HCl(aq) + H_2O(l) \rightleftharpoons H_3O^+(aq) + Cl^-(aq) \qquad (2)$$

(I) Beneath each species, write the concentration of that species before the reaction occurs. Note that water doesn't appear in the equilibrium constant expression, so water can be ignored.

(C) Write the change in concentration of each species that reacts or is produced. Note that HCl is a strong acid so essentially all of it reacts to produce an equal amount of hydronium and chloride ions as indicated by the stoichiometric coefficients in the balanced reaction equation.

(E) Find the equilibrium concentrations: subtract the change in concentration from the initial concentration for each reactant, and add the change in concentration to the initial concentration for each product.

Table I. RICE Table for Strong Acid Ionization

R) Reaction eq.	
I) Initial conc.	
C) Change in conc.	
E) Equilibrium conc.	

Finally, calculate the pH from the hydronium ion concentration that you entered in Table I.

Exploration 2: Determining pK_a from pH.

Lactic acid, $CH_3CH_2(OH)COOH$, is found in a number of foods. Find the pK_a of lactic acid given that a 0.25 M lactic acid solution has a pH of 2.23.

Complete Table II to help you solve this problem. Initially the concentration of lactic acid was 0.25 M. The acid ionization reaction produces hydronium ion, and the equilibrium concentration of the hydronium ion can be determined from the pH. Note from the reaction equation that the equilibrium concentration of the lactate anion must equal the hydronium ion concentration

Table II. RICE Table for pKa of lactic acid

R) Reaction equation	
I) Initial concentration	
C) Change in concentration	
E) Equilibrium concentration	

Finally, use the equilibrium concentrations that you find to calculate K_a and pK_a.

Exploration 3: Determining pH from K_a

Carbon dioxide dissolves in water to form carbonic acid, H_2CO_3. What is the pH of natural rainwater that is saturated with carbon dioxide? The pK_{a1} for losing the first H^+ from carbonic acid is 6.38, and the concentration of carbonic acid in a saturated solution of carbon dioxide is 1.09×10^{-5} M.

The initial amount of carbonic acid is 1.09×10^{-5} moles in 1.0 L of solution. You need to find the amount that ionizes to form hydronium ion; call this amount x. The equilibrium concentration of carbonic acid then is $(1.09 \times 10^{-5}) - x$ M. In this expression, the x actually can be neglected because K_{a1} is so small, and consequently the amount that ionizes, x, will be small compared to the amount present initially.

In situations where this approximation is not valid, the expression for the equilibrium constant produces a quadratic equation in x that can be solved using the quadratic formula.

Complete the RICE table below, then determine the pH of rainwater.

Table III. Rice Table for the pH of rainwater.

R) Reaction equation	
I) Initial conc.	
C) Change in conc.	
E) Equilibrium conc.	

Got It!

1. Why is writing the balanced reaction equation an important part of the RICE table strategy especially when some of the stoichiometric coefficients differ from 1?

2. In the problem dealing with the ionization of lactic acid, why are the hydronium ion and lactate concentrations equal?

3. In the problem dealing with the pH of rainwater, why is the amount of carbonic acid present at equilibrium taken to be the amount of carbonic acid present initially despite the fact that some of the carbonic acid reacts and ionizes?

4. What insight about solving acid equilibrium problems did you or your team gain from this activity so far?

Additional Practice Problems

1. Acetic acid (CH_3COOH, $pK_a = 4.74$) is found in vinegar at a concentration of around 1.00 M. What are the equilibrium concentrations of hydronium ion, acetate ion, and acetic acid in such a vinegar solution? What is the pH of this solution?

159

3. Ephedrine hydrochloride, which is the active ingredient in some decongestants, is a salt like ammonium chloride $NH_4^+Cl^-$. We use the formula EH^+Cl^- for this salt. The cation EH^+ has a pK_a = 9.96. The pH of blood (7.4) determines the hydronium ion concentration in the ephedrine acid ionization equilibrium.

$$EH^+(aq) + H_2O(l) \leftrightarrows H_3O^+(aq) + E(aq) \tag{3}$$

Calculate a value for the ratio $[E]/[EH^+]$ in the bloodstream, and identify which form, the acid EH^+ or its conjugate base E, is predominantly present in the bloodstream. Explain why one form dominates over the other.

Ephedrine
$C_{10}H_{15}NO$

3. Boric acid is a weak acid that often is used for an eyewash.
(a) Draw the Lewis structure of boric acid.

(b) The reaction of boric acid, $B(OH)_3$, with water is not typical acid ionization. Rather, the following reaction occurs.

$$B(OH)_3 + 2H_2O \leftrightarrows B(OH)_4^- + H_3O^+ \tag{4}$$

What feature of the boric acid Lewis structure makes this reaction possible?

(c) Calculate the pH of a 0.10 M solution of boric acid using a pK value for the reaction in equation (4) of 9.24.

16-1. Buffers: How can your blood pH be kept near the critical value of 7.4?

Many reactions and structures in biological systems, especially those involving proteins, require that a biomolecule be either in the acid form with a proton or in the base form without the proton in order to be active and functional. The ambient pH determines which form is present. If the pH is too low or too high, then the equilibrium will be shifted in the wrong direction, and life processes will cease! Consequently it is essential for biological systems to have mechanisms for controlling and maintaining the correct pH. One of these mechanisms is the use of buffers.

To understand what a buffer is and how it works, consider the equilibrium involving a weak acid and the salt of its conjugate base in an aqueous solution. The pH of this solution can be calculated from the equilibrium constant expression. For example, consider 1.0 L of solution containing 0.01 mole of NaF and 0.01 mole of HF, which has $K_a = 7.2 \times 10^{-4}$. The equilibrium constant expression for the ionization reaction

$$HF(aq) + H_2O(l) \leftrightarrows H_3O^+(aq) + F^-(aq) \qquad (1)$$

is

$$K_a = \frac{[H_3O^+][F^-]}{[HF]}. \qquad (2)$$

Since the HF and F^- concentrations change very little in establishing equilibrium and the K_a value is known, Equation (2) can be solved for the hydronium ion concentration and then the pH calculated, since $pH = -\log([H_3O^+])$

Equation (2) also can be manipulated to produce a new equation that is a bit easier to use. Take minus log of both sides and use the equality $\log(ab) = \log(a) + \log(b)$ to obtain Equation (3).

$$-\log(K_a) = -\log[H_3O^+] - \log\left(\frac{[F^-]}{[HF]}\right). \qquad (3)$$

Then replace the minus log quantities with pK_a and pH and rearrange to obtain Equation (4), which is known as the Henderson-Hasselbalch equation. Just like $pH = -\log([H_3O^+])$, $pK_a = -\log(K_a)$.

$$pH = pK_a + \log \left(\frac{[F^-]}{[HF]} \right) \tag{4}$$

Exploration

In this exploration, you will complete Table I below to show the effect of adding a strong acid or base to pure water compared to adding these substances to a solution of a weak acid and its conjugate base. Enter your answers to the following questions in Table I.

1. What is the pH of the solution when 0.0010 mole of HCl is added to make 1.0 L of solution?

2. What is the pH of the solution if 0.0010 mole of sodium hydroxide is added to make 1.0 L of solution?

3. Use equation (4) to calculate the pH of 1.0 L of solution containing 0.050 mole HF and 0.050 mole NaF.

4. Calculate the pH of the solution in *Exploration* question 3 after 0.0010 mole of hydrochloric acid, HCl, has been added to it. Construct a *RICE Table* to determine the new equilibrium concentrations of HF and F. Note that the 0.0010 mole of HCl reacts with 0.0010 mole of F^- to produce an additional 0.0010 mole of HF.

5. Calculate the pH of the solution in *Exploration* question 3 after 0.0010 mole of a strong base, sodium hydroxide NaOH, has been added to it. Note that the base converts 0.0010 mole of HF to F⁻.

Table I. Effect of a strong acid and a strong base on pH

Question	Solution	Amount	pH
	Pure water	1.0 L	7.0
1	HCl added	0.010 mole added	
2	NaOH added	0.010 mole added	
3	0.50 M in both HF and NaF	1.0 L	
4	HCl added	0.010 mole added	
5	NaOH added	0.010 mole added	

6. In view of your calculated results summarized in Table I, what can you say about the relative effects of a strong acid and a strong base on the pH of the HF/NaF solution compared to the effect on the pH of pure water?

7. Why do you think the addition of a strong acid or a strong base to the HF/NaF solution had a much smaller effect on the pH compared to the effect with pure water?

Information

A *buffer* or a *buffer solution* is a chemical system that resists changes in pH when an acid or a base is added to it or when it is diluted. A buffer solution is composed of a weak acid and its conjugate base.

A buffer resists changes in pH because the weak acid reacts with and consumes any strong base that is added and the weak acid's conjugate base reacts with and consumes any strong acid that is added.

The pH of a buffer solution is determined by the acid's pK_a value and the ratio of base/acid concentrations as given by the Henderson-Hasselbalch equation.

The effectiveness or capacity of the buffer is determined by the amounts of weak acid and its conjugate base present in the solution.

Got It!

1. According to Equation (4), what determines the pH of a buffer if the weak acid and its conjugate base are present in equal amounts?

2. Suppose you wanted to make up another solution that resisted pH change just like the HF/NaF solution did, but had a pH value close to 6. What would you do to make such a solution based on the variables or parameters in the Henderson-Hasselbalch equation?

3. Calculate the pH of a solution 0.0150 M in acetic acid (pKa = 4.74) and 0.300 M in sodium acetate.

Application

1. The pH of blood is buffered by several mechanisms to maintain a value close to 7.4. Calculate the ratio HPO_4^{2-} / $H_2PO_4^-$ that is needed to produce a buffer at pH = 7.4. ($K_a = 6.2 \times 10^{-8}$ for dihydrogen phosphate.)

2. A buffer consists of 0.30 M propanoic acid ($K_a = 1.4 \times 10^{-5}$) and 0.30 M sodium propanoate.
   (a) What is the pH of this buffer?

   (b) What is the pH after 1.0 mL of 0.10 M HCl is added to 10 mL of the buffer?

   (c) What is the pH after 10.0 mL of 0.10 M HCl is added to 10.0 mL of the buffer?

   (d) Compare your answers to parts (b) and (c). What happens to the effectiveness of the buffer as more and more acid or base is added?

3. How many grams of ammonium chloride (NH_4Cl, pK_a = 9.25) would need to be combined with 0.050 mol NH_3 to produce 500 mL of a buffer solution with a pH of 9.00?

4, Calculate the pH change when 1.0 mL of 1.0 M NaOH is added to 100 mL of the following solutions.
(a) 0.10 M acetic acid and 0.10 M sodium acetate

(b) 0.010 M acetic acid and 0.010 M sodium acetate

(c) 0.0010 M acetic acid and 0.0010 sodium acetate

(d) What can you conclude by comparing the results of your calculations in parts a − c?

16-2. Acid – Base Titrations: How much acetic acid is there in an exotic wine vinegar?

In a titration, a substance in a solution with a known concentration reacts with a second substance in another solution. The stoichiometry of the reaction is known, and the objective is to determine the amount or concentration of the second substance. The progress of the reaction is monitored by some indicator that reveals when the equivalence point has been reached. The *equivalence point* is the point in the titration when the two substances have reacted completely so that neither is present in excess.

For this activity, you might imagine that you are working in the analytical laboratory of *Nature's Own Natural Foods, Inc.* Your task is to analyze the titration of acetic acid with sodium hydroxide to determine whether or not the titration can be used reliably to measure the concentration of acetic acid in your company's exotic vinegars. To demonstrate the validity of the titration procedure, you need to calculate the pH at various points in the titration to demonstrate that existing acid-base theory is consistent with the experimental titration data.

Exploration

To explore and validate the titration procedure, consider what happens when 25.0 mL of 0.100 M acetic acid, CH_3COOH, reacts with increasing amounts of 0.050 M sodium hydroxide, NaOH. Acetic acid is a weak acid with a pK_a of 4.74. Then summarize your results in Table I at the end.

1. What is the initial pH of the acetic acid solution? The initial pH is the pH of the solution before any sodium hydroxide is added. Use the acid ionization constant expression to calculate the hydronium ion concentration and the pH of the 0.100 M acetic acid solution.

2. What is the pH of the acetic acid solution after 10.0 mL of the sodium hydroxide solution has been added to it? Sodium hydroxide reacts with acetic acid to produce acetate ions. Determine how much acetic acid remains and how much acetate has been produced, then use the Henderson-Hasselbalch equation to calculate the pH.

3. What is the pH after 25.0 mL of the sodium hydroxide solution has been added? Determine how much acetic acid remains and how much acetate has been produced, then use the Henderson-Hasselbalch equation to calculate the pH.

4. What is the pH after 45.0 mL of the sodium hydroxide solution has been added?

5. What is the pH after 50.0 mL of the sodium hydroxide solution has been added? You should see that this is the equivalence point. There were 25.0 mL x 0.100 M = 2.50 mmol of acetic acid and now 50.0 mL x 0.050 M = 2.50 mmol of sodium hydroxide have been added. Essentially all of the acetic acid has been converted into acetate ions (Ac^-). Now, the equilibrium constant expression for the hydrolysis of acetate, K_b, can be used to calculate the hydroxide ion concentration, which then can be converted to the pOH and the pH.

$$Ac^-(aq) + H_2O(l) \rightleftharpoons HAc(aq) + OH^-(aq) \qquad pK_b = 9.26$$

6. What is the pH after 55.0 mL of the sodium hydroxide solution has been added? The sodium hydroxide in 50.0 mL of solution reacted with all of the acetic acid, so now there is 5.00 mL of excess sodium hydroxide or 0.25 mmol in 80.0 mL of solution. The concentration of excess sodium hydroxide is the hydroxide ion concentration, which can be converted to pOH and pH.

7. What is the pH after 75.0 mL of the sodium hydroxide ion solution has been added? Determine the amount of sodium hydroxide in excess, and divide by the volume of the final solution to get the hydroxide ion concentration.

8. Summarize the results of you calculations from Exploration questions 1 – 7 in the following table. Identify the major species present in each solution as acetic acid, acetate, or both. The major species are the ones with the highest concentrations that determine the pH.

Table I. Calculated data for the titration of 25.0 mL of 0.100 M acetic acid with 0.050 M sodium hydroxide.

Question	Situation	Major Species	$[H_3O^+]$	pH
1	Before adding NaOH solution			
2	10.0 mL NaOH			
3	25.0 mL NaOH			
4	45.0 mL NaOH			
5	50.0 mL NaOH			
6	55.0 mL NaOH			
7	75.0 mL NaOH			

9. Draw a graph below of your data in Table I. Plot pH on the y-axis and the volume of the NaOH solution added on the x-axis. A high quality graph has a title and axes with labels and scales. Points are plotted with small circles around them to make them easy to see, and a smooth line is drawn through the data points to show how the trend continues between the points.

Got It!

The titration curves obtained for two acids being titrated with the same base are shown in the figure to the right.

1. Which curve is for the weaker acid? Explain.

2. What is the approximate pH at the equivalence point for each acid?

3. Why does the pH at the equivalence point differ for the two acids?

4. Why does the pH at the beginning of the titration differ for the two acids?

5. Why is the pH at the end of the titration (past the equivalence point) the same for both acids?

171

6. At what point in the titration of a weak acid with a strong base is the pH equal to the pK$_a$ of the acid? Explain.

7. Where in the titration of a weak acid with a strong base is a buffer solution produced? Explain.

8. Why does the pH change so abruptly near the equivalence point?

Application

1. *Nature's Own Natural Foods, Inc.*'s (NONF) premier product is champagne vinegar. Your lab technician titrated 50.0 mL of this specialty item with 0.500 M sodium hydroxide and found the equivalence point at 84.0 mL.
(a) What is the molar concentration of acetic acid in NONF's champagne vinegar?

(b) What is the mass percent of acetic acid in this product, which has a density of 1.010 g/mL? The molar mass of acetic acid is 60.05 g/mol.

(c) Is the amount of acetic acid in NONF's champagne vinegar consistent with the label, which claims 5% (by weight)? Explain.

172

17-1. Entropy: Why do some things occur spontaneously while others do not?

Every process has a preferred direction, which is called the spontaneous direction. For example, snow melts on a warm day, iron rusts when in contact with water and oxygen, and water runs downhill. A *spontaneous process* is an action that occurs on its own accord unless restrained in some way, and work must be done or energy expended in order to reverse a spontaneous process. For example, water runs downhill unless restrained by a dam, and a pump is needed to move water uphill. Some examples of spontaneous and non-spontaneous processes are given in Table I. These examples all presume a temperature of 298 K because whether a process is spontaneous or not can depend on the temperature. The change in enthalpy, ΔH, refers to whether energy is released ($\Delta H < 0$), an exothermic process) or absorbed ($\Delta H > 0$, an endothermic process).

What do you think?

Respond to the following questions using the characteristics of spontaneous and non-spontaneous processes in Table I and others that you might think of.

1. Are exothermic processes always spontaneous?

2. Are endothermic processes always spontaneous?

3. In view of the information in Table I and your experience, what do you see as common characteristics of endothermic processes that are spontaneous?

4. What do you think causes some processes to occur spontaneously?

Table I. Spontaneous and Non-spontaneous Processes at 298 K

Process at 298 K	Initial State	Final State	ΔH	Spontaneous
Iron rusting $4Fe(s)+3O_2(g)\rightarrow 2Fe_2O_3(s)$	solid & gas	solid	exo	yes
Glucose metabolism $C_6H_{12}O_6(s)+6O_2(g)\rightarrow 6CO_2(g)+6H_2O(l)$	solid & gas	liquid & gas	exo	yes
Drying agent working $CaSO_4(s)+2H_2O(g)\rightarrow CaSO_4\cdot 2H_2O(s)$	solid & gas	solid	exo	yes
Water decomposing $2H_2O(l)\rightarrow 2H_2(g)+O_2(g)$	liquid	gas	endo	no
Salt dissolving in water $NaCl(s)\rightarrow Na^+(aq)+Cl^-(aq)$	solid	ions in solution	endo	yes
Ice melting $H_2O(s)\rightarrow H_2O(l)$	solid	liquid	endo	yes, at 298 K
Water freezing $H_2O(l)\rightarrow H_2O(s)$	liquid	solid	exo	no, not at 298 K

exo = exothermic, ΔH<0, endo = endothermic, ΔH>0.

Information

A spontaneous process occurs because it produces a final situation that is much more probable than the initial situation. The most probable situation is the one that has the largest number of states accessible to the system.

To understand this idea in familiar terms, consider the possible states of books on your bookshelf. There is only one state in which all the books are lined up alphabetically by author starting on the left with the binding facing outward and the titles on the binding facing left, but there are many states if these constraints are removed. During the semester, unless you make a special effort exerting work and energy, your bookshelf will evolve spontaneously into one of these more probable states.

A spontaneous process can be either endothermic or exothermic, but whenever a process is spontaneous, the process causes energy to disperse or spread out over more states. This dispersal of energy occurs because there are more ways for the energy to be spread out over many states than there are for it to be concentrated in only one or a few states.

174

Exploration - 1

Suppose you have ten atoms and two quanta (units) of energy as shown in Figure 1. Both quanta could be used to excite one atom as shown in the first row of the figure, or each quantum could excite a different atom as shown in the second row. Consider each distinct assignment of quanta to the atoms to represent a microstate of the system. So the first row represents one microstate, and the second row represents a different microstate.

Fig. 1. Stars represent quanta of energy, *e.g.* supplied by photons. Two quanta can be used to excite one atom, or each quantum can excite a different atom.

1.1. How many microstates are there with 2 quanta of energy assigned to the same atom, i.e. how many different pictures similar to the first row could you draw?

1.2. How many microstates are there with each quantum assigned to a different atom, i.e. how many different pictures similar to the second row could you draw?

1.3. What is the probability of having an energy-dispersed state where each quantum is on a different atom? To calculate this probability, divide your answer to Question 1.2 by the sum of your answers to Questions 1.1 and 1.2.

1.4. If such a system of ten atoms and two quanta of energy were allowed to evolve spontaneously by exchanging energy from one atom to another, would you expect to find it in one of the energy-dispersed states or in one of the energy-localized states? Explain.

Information

The idea that spontaneous processes occur to disperse energy and maximize the number of accessible states is quantified in the concept of *entropy*.

Ludwig Boltzmann and Rudolf Clausius, Austrian and German physicists, made these connections in the late 19th century.

Boltzmann defined entropy by the equation

$$S = k \ln(W) \tag{1}$$

where S is the entropy, k is a fundamental constant now known as the Boltzmann constant (1.3807×10^{-23} J/K), and W is the number of accessible microstates of the system. Equation (1) shows that *entropy* is a measure of the number of accessible states and the dispersal of energy.

A *microstate* of a system is the specification of the states of all the individual atoms and molecules comprising the system. For example, a microstate of a system consisting of ten hydrogen atoms would be represented by listing the quantum numbers $n, l, m_l,$ and m_s for each hydrogen atom plus the position and momentum of each atom. The term *microstate* is used to distinguish these states from the *macrostates* of matter: solid, liquid, and gas.

Table II provides the entropy changes for the processes listed in Table I.

Table II. Change in the Number of Accessible States as Measured by Entropy

Process at 298 K	ΔS_{system} (J/K)	$\Delta S_{surround}$ (J/K)	ΔH	Spontaneous
Iron rusting $4Fe(s)+3O_2(g) \rightarrow 2Fe_2O_3(s)$	87	5532	exo	yes
Glucose metabolism $C_6H_{12}O_6(s)+6O_2(g) \rightarrow 6CO_2(g)+6H_2O(l)$	235	9396	exo	yes
Drying agent working $CaSO_4(s)+2H_2O(g) \rightarrow CaSO_4 \cdot 2H_2O(s)$	−290	352	exo	yes
Water decomposing $2H_2O(l) \rightarrow 2H_2(g)+O_2(g)$	327	−1918	endo	no
Salt dissolving in water $NaCl(s) \rightarrow Na^+(aq)+Cl^-(aq)$	43	−13	endo	yes
Ice melting $H_2O(s) \rightarrow H_2O(l)$	22.00	−20.16	endo	yes, at 298 K
Water freezing $H_2O(l) \rightarrow H_2O(s)$	−22.00	20.16	exo	no, not at 298 K

exo = exothermic, $\Delta H < 0$, endo = endothermic, $\Delta H > 0$.

Exploration - 2

2.1. Based on the data in Table II, does an exothermic process always cause the entropy of the system to increase? Explain or justify your response.

2.2. Based on the data in Table II, does an exothermic process always cause the entropy of the surroundings to increase? Explain or justify your response.

2.3. Based on the data in Table II, does the entropy of the system always increase for a spontaneous process?

2.4. Based on the data in Table II, does the entropy of the surroundings always increase for a spontaneous process?

2.5. Calculate the total entropy change ($\Delta S_{system} + \Delta S_{surroundings}$) for each process in Table II and summarize your results below. Does the total entropy always increase for a spontaneous process?

2.6. Both water decomposing and sodium chloride dissolving in water are endothermic processes. Why is one spontaneous and the other one not spontaneous?

2.7. What correlation have you found between the change in the entropy, the change in the number of accessible states, and whether a process is spontaneous or not?

Information

It was Boltzmann's idea that processes evolve spontaneously to produce the most probable situation, the one with the largest number of accessible microstates. From this perspective, the total entropy for any spontaneous process must increase. The change in the total entropy is the sum of two terms. One is the change in the entropy of the system, ΔS_{system}. The other term is the change in the entropy of the surroundings, $\Delta S_{surroundings}$. The system is defined by the molecules involved in the process or reaction. The surroundings are everything else that provides energy in an endothermic process or absorbs energy in an exothermic process. Sometimes the *total entropy* is called the *entropy of the universe.*

$$\Delta S_{universe} = \Delta S_{system} + \Delta S_{surroundings} \qquad (2)$$

The number of system states depends on the number and type of molecules and the phase. Generally, the number of states increases if the number of molecules increases, or if the phase goes from solid to liquid (or solution) or from liquid to gas.

In an exothermic process, energy is transferred to the surroundings; therefore the number of accessible states in the surroundings increases because the added energy makes higher energy states accessible. In an endothermic process, energy is transferred from the surroundings; therefore the number of accessible states in the surroundings decreases because the higher energy states no longer are accessible.

Clausius discovered that the change in entropy, ΔS, for any reversible process at a constant temperature is related to the energy, q, that is dispersed or transferred:

$$\Delta S = q/T \qquad (3)$$

where T is the Kelvin temperature.

The change in the entropy of the surroundings, $\Delta S_{surroundings}$, can be calculated from the enthalpy change for the process by using Equation (3) since the amount of energy transferred into the surroundings, $q_{surroundings}$, is given by

$$q_{surroundings} = -\Delta H_{system} \qquad (4)$$

The minus sign in this equation accounts for the inverse relationship between energy transfer with respect the system and with respect to the surroundings. For example, for energy transferred out of the system, q_{system} and ΔH_{system} are negative but $q_{surroundings}$ is positive because energy is being added to the surroundings.

Got It!

Whether or not a process is spontaneous or not often depends on the temperature. A common example is water freezing and ice melting. Water freezes spontaneously only if the temperature is below 273 K, and ice melts spontaneously only if the temperature is above 273 K. At 273 K water and ice are in equilibrium, which means that the amount of ice and water present does not change.

1. In terms of the entropy changes in Table II, why does ice melt spontaneously at 298 K rather than water freezing at this temperature?

2. The enthalpy of fusion of water is 6009 J/mol, and the entropy of fusion is 22.0 J/K mol. Use Equations 2, 3, and 4 to determine the temperature at which the total entropy change is zero when (a) one mole of ice melts and (b) one mole of water freezes.

3. At what temperatures will water freeze spontaneously? Explain in terms of the entropy changes of the system and the surroundings.

Information

When ice melts, the entropy and number of accessible states of the system increase (molecules have fixed positions in the solid but many positions and velocities in the liquid), but this process is endothermic, so the entropy and number of accessible states of the surroundings decrease. The effect that dominates depends on the temperature.

The surroundings are not important at high temperatures because at high temperatures the surroundings have much energy and many accessible states. Consequently at high temperatures (above 273 K), the entropy increase of the system dominates and ice melts spontaneously.

At low temperatures (below 273 K), the entropy change in the surroundings is more important, so the exothermic process (water freezing) is spontaneous because it increases the entropy and number of accessible states in the surroundings.

The melting/freezing point of water is 273 K because at this temperature, the contributions to the entropy change from the system and the surroundings cancel, giving no change in the total entropy and total number of accessible states. Consequently, water and ice are in equilibrium at this temperature, and there is no spontaneous change in the amount of water and ice present.

Application

1. Water was put into a freezer and spontaneously froze at 0°C to produce 9.0 g of ice (0.5 mol). The temperature of the freezer was −15°C (258 K). The enthalpy of fusion of ice is 6.01 kJ/mol.

    (a) When the water froze, did energy flow from the water to the freezer or from the freezer to the water?

    (b) Use equation (3) to calculate a value for the change in entropy of the system (water and ice). Note that q_{system} is negative if energy flowed into the surroundings and positive if energy flowed into the system, and that the water froze at 273 K.

    (c) Calculate a value for the change in entropy of the surroundings, i.e. the freezer at a temperature of 258 K.

    (d) Explain why the water froze spontaneously when it was put into the freezer in terms of the change in the number of accessible states and the entropy of the system (water and ice) and the surroundings (the freezer).

17-2. Free Energy: How much work can be obtained from any process?

What do you think?

Can a process be spontaneous at one temperature but not at another? Explain or provide an example to support your answer.

Information

The entropy change of a chemical reaction, $\Delta S_{reaction}$, depends on the difference in the entropy of the products and reactants.

$$\Delta S_{reaction} = S_{products} - S_{reactants} \qquad (1)$$

The entropy of the surroundings, $\Delta S_{surroundings}$, also changes as a result of a chemical reaction because energy is transferred either into the surroundings from the system (an exothermic reaction) or into the system from the surroundings (an endothermic reaction).

A chemical reaction is spontaneous if it causes an increase in the entropy of the universe, $\Delta S_{universe}$, where

$$\Delta S_{universe} = \Delta S_{surroundings} + \Delta S_{reaction} \qquad (2)$$

An exothermic reaction (enthalpy change, $\Delta H_{reaction} < 0$) transfers energy to the surroundings, thereby making more states accessible, and increasing the entropy. An endothermic reaction ($\Delta H_{reaction} > 0$) removes energy from the surroundings, thereby reducing the number of accessible states, and decreasing the entropy. The entropy change of the surroundings can be calculated from Clausius's definition $\Delta S = q/T$ by using equation (3).

$$\Delta S_{surroundings} = - \Delta H_{reaction} / T \qquad (3)$$

Combining equations (2) and (3) gives equation (4)

$$\Delta S_{universe} = - \Delta H_{reaction} / T + \Delta S_{reaction} \qquad (4)$$

The right-hand side of equation (4) depends only on properties of the reaction: the enthalpy and entropy changes and the Kelvin temperature. Rather than call this quantity $\Delta S_{universe}$, J. Willard Gibbs defined a new quantity that we now call the *Gibbs free energy*, G. Gibbs decided to define the change in free energy as minus T times the change in the entropy of the universe,

$$\Delta G_{reaction} = - T \Delta S_{universe}. \qquad (5)$$

With this definition, equation (4) becomes

$$\Delta G_{reaction} = \Delta H_{reaction} - T \, \Delta S_{reaction} \qquad (6)$$

Since each quantity in equation (6) has the subscript *reaction*, it is not necessary to continue to include the subscript.

Although the situation for chemical reactions is considered explicitly in the above discussion, the conclusion can be applied to any process, and ΔG is called the free energy because the decrease in this quantity corresponds to the maximum amount of work that can be obtained from any process at constant pressure.

$$\Delta G = \Delta H - T \, \Delta S \qquad (7)$$

Exploration

1. What quantity gives the amount of energy that is released to the surroundings or absorbed from the surroundings by a chemical reaction?

2. In equation (3), why is there a minus sign in the equation, $-\Delta H_{reaction}$ and not just $\Delta H_{reaction}$?

3. Given that ΔG for the reaction $2H_2O(g) \rightarrow 2H_2(g) + O_2(g)$ is positive, which reaction is spontaneous: the formation of water from hydrogen and oxygen or the decomposition of water to produce hydrogen and oxygen? Explain in terms of the sign of ΔG and the change in the entropy of the universe.

4. If four processes happen to have the same magnitudes for the enthalpy change and for the entropy change but different signs for these quantities, which one is capable of accomplishing the most work? Explain.

   (a) $\Delta H>0$, $\Delta S>0$ (b) $\Delta H>0$, $\Delta S<0$ (c) $\Delta H<0$, $\Delta S>0$ (d) $\Delta H<$, $\Delta S<0$

5. Why, in terms of equation (7), might a reaction be spontaneous at one temperature but not at another?

Got It!

Some reactions are spontaneous at any temperature, others are not spontaneous at any temperature, some are spontaneous only at low temperatures, and others are spontaneous only at high temperatures. These four characteristics are a consequence of four combinations of ΔH and ΔS that are possible. Complete the following table by identifying whether ΔG will be positive or negative and which of the above 4 characteristics result from the combination of ΔH and ΔS given for each row of the table.

Table I. Effects of ΔH, ΔS, and temperature on reaction spontaneity.

ΔH	ΔS	Sign of ΔG $\Delta G = \Delta H - T\Delta S$	Spontaneous always, never, at low T, or at high T?
> 0	> 0		
> 0	< 0		
< 0	> 0		
< 0	< 0		

Application

1. Chemical reactions can be used in hot packs and cold packs for treating athletic injuries. Consider the following reaction, and using information about standard enthalpies, entropies, and free energies found in your textbook or elsewhere answer the questions below.

$$NH_4NO_3(s) \rightarrow NH_4^+(aq) + NO_3^-(aq)$$

(a) Is this reaction spontaneous or not at room temperature? Explain.

(b) Would this reaction be useful or not in a hot pack or a cold pack? Explain.

2. You are applying for a job in an analytical lab at a silver mine in the beautiful Black Hills of South Dakota. On the application form, they ask you to determine the minimum temperature to which silver oxide must be heated in order to produce silver metal from the ore. What is your answer? You are provided with the following information.

$$2Ag_2O(s) \rightarrow 4Ag(s) + O_2(g)$$

Quantity	$Ag_2O(s)$	$Ag(s)$	$O_2(g)$
ΔG_f° (kJ/mol)	-11.7	0	0
ΔH_f° (kJ/mol)	-31.1	0	0
S° (kJ/mol)	121	43	205

18-1. Electrochemical Cells: Where does the energy come from to power your cell phone?

Batteries supply electrical current to power various devices that you use. On the street and in Radio Shack usually no distinction is made between an electrochemical cell and a battery, but actually a battery consists of several electrochemical cells connected together to produce a higher voltage. For example, a 12 volt automobile battery consists of six 2 volt cells connected together. An electrochemical cell also is called a *voltaic cell* or a *galvanic cell*.

What do you think?

Batteries power various devices that you use. What do you think is the source of the electrical power from a battery?

(A) Electrons stored on a metal film inside the battery.

(B) Ions moving from one end of the battery to the other through the external circuit.

(C) Electrons produced by one chemical reaction are consumed in another reaction.

(D) Electrons produced by a chemical reaction are consumed in powering the device.

Exploration

1. Identify the components of a Zn/Cu electrochemical cell from Figure 1 and complete the table below.

Component	Anode Compartment	Cathode Compartment
Electrode metal (Cu or Zn)		
Electron potential (+ or -)		
Type of half reaction		
Write the half reaction		

$E^\circ_{cell} = +1.10\ V$

$E^\circ_{ox} = +0.76\ V$

Anode ⊖

K₂SO₄ salt bridge

Cathode ⊕

$E^\circ_{red} = +0.34\ V$

Anode compartment

SO_4^{2-} ← → K^+

Cathode compartment

$Zn \longrightarrow Zn^{2+} + 2\ e^-$
oxidation

Porous plugs

$Cu^{2+} + 2\ e^- \longrightarrow Cu$
reduction

Zn^{2+}

Cu^{2+}

Net reaction:

$Cu^{2+}(aq) + Zn(s) \longrightarrow Cu(s) + Zn^{2+}(aq)$

Fig. 1. An electrochemical cell being used to light a light bulb. This cell has Zn and Cu metal electrodes immersed in 1.0 M ZnSO₄ and CuSO₄ solutions, respectively.

2. Write the reaction that produces electrons to power the light bulb.

3. Write the reaction that consumes the electrons.

4. Which way do the electrons flow in the external circuit, from the anode to the cathode or from the cathode to the anode?

5. Is the concentration of cations in the anode compartment increasing or decreasing?

6. In view of your answer to Exploration question (5), which way do positive and negative ions flow in the salt bridge? Explain why.

7. What is the difference between an oxidation reaction and a reduction reaction?

8. The voltage produced by an electrochemical cell is called the cell potential, E_{cell}. Examine Fig. 1, and identify how the standard cell potential, E^o_{cell} of +1.10 volts, is determined from the standard half-cell potentials for the oxidation, E^o_{ox}, and reduction, E^o_{red} reactions.

9. Read the information section below, and then write an equation that you can use to calculate a standard cell potential from the standard reduction half-cell potentials.

Information

The standard cell potential, E^o_{cell}, is a measure of the force that attracts electrons to the positive electrode. The total driving force for the cell is determined by the sum of the driving forces for the oxidation reaction and for the reduction reaction. The cell potential therefore is determined by the half-cell potentials for these two reactions, E^o_{ox} and E^o_{red}.

By convention, reference tables list standard reduction potentials and the corresponding reduction reactions. The half-reaction with the more positive reduction potential has the greater force to attract electrons. For example,

$$Cu^{2+}(1M) + 2e^- \rightarrow Cu(s) \qquad E^o_{red} = + 0.34 \text{ V}$$
$$Zn^{2+}(1M) + 2e^- \rightarrow Zn(s) \qquad E^o_{red} = - 0.76 \text{ V}$$

Since the copper reaction has the more positive reduction potential, it runs as a reduction and the zinc reaction reverses and runs as an oxidation. When a half reaction is reversed, the value of the half-cell potential remains the same, but the sign reverses, so for a given half reaction, $E^o_{ox} = - E^o_{red}$, as shown by the above example.

Got It!

A silver oxide battery, which is commonly used in watches, is based on the following half reactions.

$$Ag_2O(s) + H_2O(l) + 2e^- \rightarrow 2Ag(s) + 2OH^-(aq) \qquad E^\circ_{red} = +0.34 \text{ V}$$
$$ZnO(s) + H_2O(l) + 2e^- \rightarrow Zn(s) + 2OH^-(aq) \qquad E^\circ_{red} = -1.22 \text{ V}$$

1. Which reaction will run as a reduction and which one will run as an oxidation in the silver oxide battery?

2. What is the standard cell potential of the silver oxide battery?

3. Write the reactions occurring at the anode and at the cathode and label them as *oxidation* or *reduction*.
   anode reaction:

   cathode reaction:

4. Write the overall reaction for the silver oxide battery. You add the reduction reaction and the oxidation reaction together to get the overall reaction.

5. Identify the species being oxidized and the species being reduced in the overall reaction.

6. Identify *the oxidizing agent* and the *reducing agent* in the overall reaction.

7. Identify which of the following does not occur in the silver oxide battery.
   (a) Electrons flow from $Zn(s)$ to $Ag_2O(s)$.
   (b) Silver metal is produced as the battery operates.
   (c) Corrosive hydroxide ion (OH^-) is produced.
   (d) Ions move from one electrode compartment to the other.

18-2. Electrolytic Cells: How can silver be deposited on nickel to make a beautiful friendship ring?

An electrochemical cell uses the energy derived from chemical reactions to produce electrical energy. An electrolytic cell, on the other hand, uses electrical energy to drive chemical reactions. Electrolysis is a very important industrial process. It is used to refine and purify many metals, like copper and aluminum, and it is used to plate thin coatings of one metal on top of another. An example of electrolysis, which is shown in Figure 1, is the decomposition of sodium chloride to produce chlorine gas and sodium metal.

What do you think?

Compare Figure 1 below with the figure in the previous activity on electrochemical cells. What do you see as the two most significant differences between an electrochemical and an electrolytic cell?

Fig. 1. Electrolysis of molten sodium chloride.

189

Exploration - 1

1.1. Write the half reaction occurring at the anode in Figure 1.

1.2. Write the half reaction occurring at the cathode in Figure 1.

1.3. Write the overall reaction for the electrolysis of sodium chloride.

1.4. What are the standard reduction potentials for Cl_2 and Na^+? Use your textbook or other resource to find these values.

1.5. Use your answer to Exploration question (1.4) to determine the minimum voltage that that is needed for the electrolysis of sodium chloride under standard conditions. Just like you did for an electrochemical cell, you add the standard potentials for the oxidation reaction and the reduction reaction together.

1.6. Why do electrochemical and electrolytic cells involving the same half reactions have standard cell potentials with the same magnitude but different signs?

Exploration – 2

The amount of electrical current that flows through an electrochemical cell and the change in the amount of reactants and products present are related. Exploration questions 2.1 and 2.2 will help you see how these quantities are related by determining the number of moles of sodium and chlorine that will be produced if an electrical current of 10 amperes flows through the electrolytic cell in Figure 1 for 10 hours.

2.1. Use the balanced reaction equation to determine the number of moles of sodium, Na, and the number of moles of chlorine gas, Cl_2, that are produced by the transfer of 1 mole of electrons.

2.2 Use the following information to determine the number of moles of electrons that are transferred by 10 A of current flowing for 10 hours.
   A = ampere = unit of electrical current
   C = coulomb = unit of electrical charge
      1 A = 1 C/s
   F = Faraday's constant = charge of 1 mole of electrons
      F = 96,485 C/mol
   charge transferred = current x time (1 A x 1 s = 1 C)
   moles transferred = charge transferred / Faraday's constant

2.3. Using your results for Exploration questions (2.1) and (2.2), determine the amount in moles of sodium metal and chlorine gas produced by the electrolysis of sodium chloride using 10 A for 10 hours.

Got It!

From one perspective, hydrogen obtained from water is an ideal fuel. A photovoltaic cell can convert solar energy (sunlight) into electricity. The electricity can be used to electrolyze water to produce hydrogen and oxygen. When the hydrogen and oxygen recombine, water is produced and the energy is released for use. By these means, the solar energy can be stored, transported, and recovered. The whole process is self-sustaining and non-polluting.

The electrolysis of water involves the following two half reactions.

$$2H_3O^+(aq) + 2e^- \rightarrow H_2(g) + 2H_2O(l)$$

$$O_2(g) + 4H_3O^+(aq) + 4e^- \rightarrow 6H_2O(l)$$

1. Write the overall balanced reaction equation for the electrolysis of water.

2. Identify the species being oxidized and the species being reduced.

3. Determine the minimum voltage that would be required to electrolyze water under standard conditions using data in a table of Standard Reduction Potentials.

4. Determine the time it will take to produce 22.4 L of hydrogen gas, H_2, at STP using a current of 1.00 A.

Application

1. In making a friendship ring, how long will it take to deposit 1.00 g of silver from a silver nitrate solution using a current of 1.00 A?

19-1. Radioactivity: Are there different kinds of radioactivity?

Only some numbers of protons and neutrons in the nucleus of an atom are stable. The nuclei that are unstable spontaneously transform or decay into more stable nuclei by emitting particles or electromagnetic radiation. This process is called *radioactivity* and the unstable atoms are called *radioactive*. Some of these decay pathways are listed in Table I. They involve the emission of *alpha*, *beta*, and *positron* particles plus electromagnetic radiation called *gamma rays*. An alpha particle is the nucleus of a helium atom, which consists of 2 protons and 2 neutrons. A beta particle is an electron, and a positron is a particle with the same mass as an electron but with a charge of $+1.602 \times 10^{-19}$ C rather than the electron charge of -1.602×10^{-19} C.

What do you think?

Radioactivity is both useful and dangerous. Identify three beneficial uses of radioactive substances and one reason why radioactivity can be dangerous.

Exploration - 1

Table I. Nuclear Decay Processes

Process	Symbol	ΔZ	ΔA	Example
Alpha particle emission	α			$^{234}_{92}U \rightarrow {}^{230}_{90}Th + \alpha$
Beta particle emission	β			$^{239}_{92}U \rightarrow {}^{239}_{93}Np + \beta$
Gamma ray emission	γ			$^{57}_{26}Fe^* \rightarrow {}^{57}_{26}Fe + \gamma$ (* = excited state)
Positron emission	$^{0}_{1}e$			$^{207}_{84}Po \rightarrow {}^{207}_{83}Bi + {}^{0}_{1}e$
Electron capture	$^{0}_{-1}e$			$^{7}_{4}Be + {}^{0}_{-1}e \rightarrow {}^{7}_{3}Li$

Exploration

1. Review atomic symbol notation, and identify what the left superscript and the left subscript tell you, for example in $^{234}_{92}U$.

2. The atomic symbol notation for a positron is $^{0}_{1}e$. Write the atomic symbol notation with a left superscript and a left subscript for an alpha particle, which is a helium nucleus, and for a beta particle, which is an electron.

3. Complete Table I by adding the entries for ΔZ and ΔA, where Z is the atomic number and A is the mass number

4. What is the relationship between the mass numbers of the reactants and the mass numbers of the products in a nuclear decay process?

5. What is the relationship between the atomic numbers of the reactants and the atomic numbers of the products in a nuclear decay process?

6. Table I contains examples where a proton appears to turn into a neutron and a neutron appears to turn into a proton. What particles must be emitted by a nucleus for these transformations to occur?

    proton $\rightarrow$ neutron:

    neutron $\rightarrow$ proton:

Got It!

1. Write an equation describing the radioactive decay of each of the following isotopes. The type of decay is given in parentheses.

    $^{18}_{9}F$ (*positron emission*)

    $^{26}_{13}Al$ (*electron capture*)

    $^{35}_{16}S$ (*beta emission*)

    $^{218}_{84}Po$ (*alpha emission*)

    $^{119}_{50}Sn^{*}$ (*gamma emission*)

19-2. Rates of Nuclear Decay: How long does a radioactive substance last?

What Do You Remember?

Review the chapter in your textbook on kinetics and the rates of reactions, and complete Table I. You should know how the integrated rate law combined with a graph can be used to identify the order of a reaction and determine the rate constant.

Table I. Rate Laws for Reactions

Reaction Order	Rate Law	Integrated Rate Law
0	$d[N]/dt =$	
1	$d[N]/dt =$	
2	$d[N]/dt =$	

Information

Radioactive iodine-131 commonly is used as part of the treatment for thyroid cancer because thyroid cells are unique in absorbing and concentrating iodine from the bloodstream. Thyroid cells need iodine for the synthesis of the thyroid hormones thyroxine and tri-iodothyronine.

Fig. 1   thyroxine   tri-iodothyronine

Iodine-131 decays by beta emission, and a typical treatment for thyroid cancer initially produces about 400×10^6 decay events per second. The level of radiation decreases with time as shown by the data in Figure 2 and Table II.

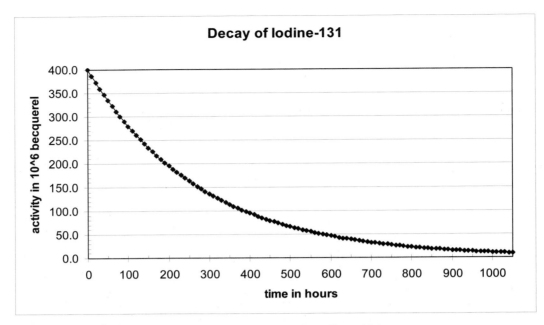

Fig. 2. Decay of Iodine-131.

Radioactivity is measured in units of decay events per second. The SI unit is the Becquerel (Bq), and 1 Bq = 1 s^{-1}. The becquerel is derived from the name of Henri Becquerel, who shared the Nobel prize with Pierre and Marie Curie for discovering radioactivity. Another commonly used unit for radioactivity is the curie (Ci), and 1 Ci = 3.7 x 10^8 s^{-1}, which is the number of decay events produced by 1 gram of radium-226.

Table II. Decay of Iodine-131

Time (hours)	Radioactivity (10^6 Bq)	Time (hours)	Radioactivity (10^6 Bq)
0	400	550	56
50	334	600	46
100	279	650	39
150	233	700	32
200	195	750	27
250	163	800	23
300	136	850	19
350	114	900	16
400	95	950	13
450	80	1000	11
500	66	1050	9

Exploration

1. Write the nuclear reaction equation for the decay of iodine-131 by beta particle emission.

2. Use the information in Figure 2 and Table II to determine whether the decay of iodine-131 is a zero, first, or second order reaction. State your conclusion and explain how it follows from the data. The information that you entered in Table I may be helpful.

3. Use the information in Figure 2 and Table II to determine
   (a) the rate constant for the decay,

   (b) the half life of iodine-131 in days.

4. Explain how your responses to Exploration Question (3) were obtained from the data.

5. Does it matter whether you write the rate law in terms of the concentration of iodine-131 or in terms of the total number of iondine-131 atoms present?

Got It!

1. As a doctor planning a treatment you need to determine how many grams of iodine-131 must be absorbed by the thyroid to produce an initial activity of 400×10^6 Bq. The rate law that you entered in Table I will be helpful to you. How many grams are needed?

2. As a doctor you need to advise a patient on how long to refrain from certain activities until the radiation level subsides. How many days does it take for the radioactivity to fall to 4×10^6 Bq from the initial level of 400×10^6 Bq?

3. If the dose were doubled to produce an initial activity of 800×10^6 Bq, how many days would it take for the activity to fall to 4×10^6 Bq?

4. A patient had an initial dose of iodine-131 that produced an activity in the thyroid of 30 mCi (millicuries). At the most recent checkup, the activity was determined to be 1 mCi. How many days had elapsed since the initial dose was administered?

20-1. Chemistry of the Main Group Elements: Why are some elements more special than others?

The Main Group Elements are those elements in Groups IA through 8A of the Periodic Table (columns 1, 2 and 13 through 18). Use your textbook and the internet as resources to explore the properties and applications of these elements.

What do you know already?

Identify 5 main group elements that you are familiar with, and identify 2 uses for each of them.

Exploration

1. How many Main Group elements are there?

2. Which 5 elements are the most abundant on Earth?

3. What are 3 methods of extracting pure elements from their ores? Provide an example of each method.

4. Identify a use for one element in each group (IA through 8A). Do not pick one of the elements that you described in the *What do you know already?* section.

5. Which 5 elements do you see as the most essential for life? Explain.

6. Which 5 elements do you see as the most useful in technology and industry? Explain.

20-2. Electronic Structure and Properties: How are electronic structure and atomic properties related?

The properties of the Main Group elements depend on their electronic structures or electron configurations. In this activity you reflect on what you have learned previously about the periodic properties of the elements, and you identify the connection between the electron configuration of an element and its properties.

The *three key ideas* to use in your responses to the questions below are the shell structure of the atom, the attraction of an electron by the positive nucleus, and the repulsion of an electron by the other electrons.

What do you remember?

Describe why atomic radii decrease across a Period and increase down a Group in the Periodic Table in terms of the three key ideas.

Exploration

1. Which Main Group elements are metals, and which are nonmetals?

2. Compared to metals, do nonmetals have high or low ionization energies? Explain why in terms of the three key ideas.

3. Compared to metals, do nonmetals have high or low electron affinities? Explain why in terms of the three key ideas.

4. Compared to metals, do nonmetals have high or low electronegativities? Explain why in terms of the three key ideas.

5. For one element in each group (IA through 8A), identify a characteristic physical or chemical property of that element and explain the origin of that property in terms of one or more of the following: its electron configuration, atomic radius, ionization energy, electron affinity, and electronegativity.

21-1. Transition Metals and Coordination Compounds: What's so special about transition metals?

The block of elements in the Periodic Table (columns 3 – 12) that have their origin in electrons filling the d-orbitals are called the *transition metals* or *transition elements*. Transition metals have many diverse and important properties. Some are very reactive and catalyze chemical reactions; others are inert and are used for coins and jewelry.

What do you know already?

Identify 5 transition metals that you are familiar with, and identify 2 uses for each of them.

Information

Many of the chemical properties of transition metals are related to their ability to form *coordinate covalent bonds* with molecules or ions that have lone pair electrons. Transition metals use d orbitals as well as s and p orbitals to accept these electrons and form a chemical bond. Consequently, the bonding structure can involve more than the 8 electrons, which is characteristic of the octet rule.

Fig. 1 Hexaquanickel(II) cation

203

Figure 1 shows the *complex ion* hexaaquanickel(II). It is formed by 6 water molecules with coordinate covalent bonds to a nickel +2 ion, which is the cation in the *coordination compound* hexaaquanickel(II) chloride, $[Ni(H_2O)_6]Cl_2$. The molecules or ions bonded to the central metal ion are called *ligands*.

Other coordination compounds and their components are listed in Table I, where CN = coordination number, Ox = oxidation number of the metal, Charge L+M = total charge on the ligands plus metal.

Table I. Examples of Coordination Compounds

Coordination Compound	Complex Ion	Ligand	Ligand Geometry	Counter Ion	CN	Charge L+M	Ox
$[Ni(H_2O)_6]Cl_2$	$[Ni(H_2O)_6]^{2+}$	H_2O	Octahedral	Cl^-	6	+2	+2
$K_3[Fe(CN)_6]$	$[Fe(CN)_6]^{3-}$					−3	+3
$[Co(NH_3)_3(OH)_3]$	None						
$Na_2[HgI_4]$	$[HgI_4]^{2-}$		Tetrahedral		4		

Exploration

1. Using the example of hexaaquanickel(II) chloride in Table I, identify the relationship between the total charge on the ligands plus metal (which is the *complex*) and the total charge on the counterions.

2. Using the example of hexaaquanickel(II) chloride in Table I, identify the relationship between the total charge on the ligands plus metal (*complex*), the total charge on the ligands, and the oxidation number of the metal.

3. Using the information in Table I, identify the geometries of coordination compounds or ions with 4 and 6 ligands.

4. Complete the missing entries in Table I.

5. What components (metal, ligands, counter-ions) of a coordination compound are present in all such compounds?

6. What additional component may be present in some coordination compounds?

7. What purpose is served by the square bracket in the chemical formula for a coordination compound?

8. When an ionic coordination compound dissolves in water, it dissociates into the complex ion and the counter-ions. What are the dissociation products for each coordination compound in Table I? Which coordination compound in Table I is not ionic?

Got It!

1. Based on what you have seen in this activity, write definitions of the following terms in your own words.

   coordinate covalent bond

   coordination compound

   complex ion

   ligand

   coordination number

   counter-ion

Research

   Use your textbook and the internet as resources to explore the properties and applications of transition metals and answer the following questions.

1. Identify 5 transition metals that are active in biological systems and describe the function of each.

2. Identify 5 transition metals or coordination compounds that are useful in technology and industry and describe the function of each.

21-2. Magnetism and Color in Coordination Compounds: Where does all that attraction and beauty come from?

Brilliant colors and magnetism are characteristic of many coordination compounds. Ligand-field theory (aka crystal-field theory) is a simple model that explains these properties. In ligand-field theory, the bonding between the ligands and the metal ion is considered to be electrostatic. The negative electrons on the ligands are attracted to the positive metal ion. The presence of the ligands alters the energy of the metal's d-orbitals. Unpaired electrons in the d-orbitals produce magnetism, and electronic transitions between the d-orbitals absorb light and cause the brilliant colors.

What do you remember?

Use the Aufbau Principle, Pauli Exclusion Principle, and Hunds Rule to predict which transition metals in Period 4 (Sc through Zn) have unpaired electrons.

Exploration – 1

According to the ligand-field model, the energies of the d-orbitals are a function of the relative positions of the ligands and the d-orbitals as shown in Figure 1.

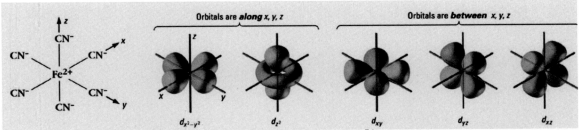

Fig. 1. The spatial arrangement of d-orbitals of an Fe^{2+} ion and the octahedral configuration of 6 CN^- ligands. The ligands are aligned along the x, y, and z axes and interact most strongly with electrons in the 2 orbitals that also are aligned along the axes and not so strongly with electrons in the 3 orbitals that are aligned between the axes.

1.1. Which d-orbitals lie along the x, y, or z axes, and which lie between any two axes?

1.2. Which orbitals experience a direct head-on interaction with the ligands that also lie along the axes?

1.3. Which orbitals do not experience a direct head-on interaction with the ligands?

1.4. Based on the nature of the interactions that you identified above in Exploration Questions 1.2 and 1.3, divide the five d-orbitals into two sets, and identify the set that has the stronger interactions with the ligands.

Set 1:

Set 2:

1.5. In a coordination compound with octahedral geometry, the energies of the two sets of orbitals depend upon the strength of the interactions with the ligands. The set experiencing the stronger interactions will be higher in energy because the electron-electron interactions are repulsive. Draw an energy-level diagram below to show the relative energies of the two sets of orbitals that you identified in Exploration Question 1.4. Represent each of the 5 orbitals by a line along with the label for that orbital.

Exploration – 2

The diagram that you drew in response to Exploration Question 1.5 is called a *ligand-field-splitting energy-level diagram*. The energy difference between the two sets of orbitals is referred to as Δ_o, where o refers to the octahedral geometry. Δ_o is called the *ligand-field* or *crystal-field* splitting, since the first research was conducted on transition metals in crystals.

2.1. Which ligand shown in Figure 2 produces the larger ligand-field splitting?

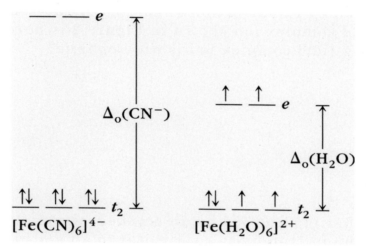

**Fig. 2. Ligand- field splitting for cyanide and water.
Cyanide is a strong field ligand and water is a weak field ligand**

2.2. What is the difference between the ways electrons fill d-orbitals in the weak-field complex ion and in the strong-field complex ion shown in Figure 2?

2.3. Since electrons fill orbitals to produce the lowest energy state, why do you think electrons are unpaired in the free ion and in the weak-field complex ion, but pair up in the strong-field complex ion? Use the relative magnitudes of the ligand-field splitting Δ_o and the electron-electron repulsion energy in your answer.

2.4. Why do you think weak-field complex ions sometimes are called high spin, and strong-field complex ions sometimes are called low spin?

2.5. An electron acts like a tiny bar magnet with the polarity determined by the direction of spin. Why do you think the weak-field complex ion shown in Figure 2 is *paramagnetic* and the strong-field complex ion is *diamagnetic*?

Exploration – 3

Light that is absorbed by an octahedral complex can be used to promote an electron from a t_2-orbital to an e-orbital. To do so, the energy in the photon, $h\nu$ or hc/λ, must equal the ligand-field splitting, Δ_o. Table I gives the approximate wavelength of light that is absorbed by 4 different cobalt complex ions.

Table I. Colors of solutions of Co^{3+} complexes.

Complex	Observed	Absorbed	Wavelength (nm)	Δ_o (J)
$[Co(NH_3)_6]^{3+}$	Yellow	Bluish violet	430	
$[Co(NH_3)_5NCS]^{2+}$	Orange	Bluish green	470	
$[Co(NH_3)_5H_2O]^{3+}$	Red	Green	500	
$[Co(NH_3)_5Cl]^{2+}$	Reddish purple	Yellowish green	522	

3.1. When you look through a solution, why is the color of the solution different from the color of the light absorbed by it, as indicated in Table I?

3.2. Using the relationship between wavelength and photon energy, arrange the cobalt complexes in order of increasing energy of the photons absorbed.

3.3. Determine the value for the ligand field splitting from the energy of the photon absorbed for each of the cobalt complex ions in Table I, and add this information to the last column in the table.

3.4. Draw d-orbital energy level diagrams for the 4 cobalt complex ions in Table I to show how the ligand-field splitting changes.

Got It!

Fe^{3+} forms octahedral complexes with NCS$^-$ and NO$_2^-$. One displays greater paramagnetism than the other.

1. Write the chemical formula for each of these two complex ions.

2. Use the *spectrochemical series* to predict which complex ion is high spin and which is low spin.

3. Identify which complex ion is more paramagnetic. Explain.

4. Draw the d-orbital ligand-field splitting diagram showing which orbitals are occupied by electrons for each complex ion.

5. Predict which complex ion will absorb light with the longer wavelength. Explain.